OCEAN WAVE MODELING

OCEAN WAVE MODELING

The SWAMP Group

SPRINGER SCIENCE+BUSINESS MEDIA, LLC

Library of Congress Cataloging in Publication Data

Main entry under title:

Ocean wave modeling.

"Principal results of a wave model intercomparison study conducted by the Sea Wave Modeling Project (SWAMP) and first presented at a symposium on Wave Dynamics and Radio Probing of the Ocean Surface, held May 13-20, 1981, in Miami, Florida"—T.p. verso.
Bibliography: p.
Includes index.
1. Ocean waves—Simulation methods. 2. Sea Wave Modeling Project. I. SWAMP Group. II. Sea Wave Modeling Project.
GC213.5.O25 1985 551.47′022′0724 84-24896
ISBN 978-1-4757-6057-6 ISBN 978-1-4757-6055-2 (eBook)
DOI 10.1007/978-1-4757-6055-2

Principal results of a wave model intercomparison study conducted by the Sea Wave Modeling Project (SWAMP) and first presented at a symposium on Wave Dynamics and Radio Probing of the Ocean Surface, held May 13-20, 1981, in Miami, Florida

Originally published by Plenum Press, New York in 1985

Contents

I. The Sea Wave Modeling Project (SWAMP): Principal Results and Conclusions

The SWAMP Group

II. The SWAMP Models

I

The Sea Wave Modeling Project (SWAMP): Principal Results and Conclusions

The SWAMP Group*

* The members of the SWAMP group are listed in Appendix A.

Foreword

Early in 1979, a group of wave researchers proposed a wave model intercomparison study to clarify the interrelations existing among the various wave models which have been developed in past years for real-time wave forecasting, wave statistics compilations, or hindcast case studies. The idea was immediately welcomed by the wave modeling community, and, finally, nine wave modeling groups from the United States, Japan, and Europe participated in the exercise. The principal results of this work are presented here jointly by the Sea Wave Modeling Project (SWAMP) Group (the members of which are listed in Appendix A). Descriptions of the models used in the study are given in Part II of this volume. A more complete documentation of the entire set of numerical experiments is given in Part 2 of the Sea Wave Modeling Project (SWAMP group, 1982).

The main purpose of the intercomparison study was to test our present understanding of the physics of wind-generated surface waves from the viewpoint of wave modeling. Specifically, we wished to clarify the basic interdependence between understanding the physics of surface waves, representing the physics numerically, and predicting quantitatively the detailed space–time evolution of a two-dimensional surface wave spectrum for a given wind field. It was not our intent to carry out a model competition. In this sense there were no winners or losers: all models could claim specific strong points, and all displayed weaknesses in some areas. Some problems were common to all models. These were particularly useful in pointing out the directions required for future experimental, theoretical, or numerical work.

A further valuable aspect of the exercise (which had not been fully anticipated when the project was planned) was the detailed insight gained by the individual modeling groups into the internal intricacies of their own models. The joint discussion of the behavior of different models under the severe test case conditions chosen for the intercomparison study frequently revealed peculiar response characteristics of individual models which had remained undetected under normal operating conditions.

The immediate motivation of the intercomparison study was the marked expansion in the operational application of wave models in recent years. This has followed a fundamental change in our view of the wave spectral energy balance in the early 1970s, when the dominant role of nonlinear interactions in controlling the growth of wind waves became apparent. One of the principal concerns of the study was therefore to clarify the differences between the first-generation wave models developed in the 1960s and the second-generation models based on the changed physical picture in the 1970s. However, significant differences were also to be expected between the various realizations of second-generation models, both with respect to the parameterization of physical processes (in particular nonlinear energy transfer) and with regard to numerical implementation.

Even for a specialist it is often difficult to distinguish between effects due to the more reliable features of a wave model, which are supported by theory and observations, and the less reliable properties, representing extrapolations of observations or parameterizations. This applies particularly to wave predictions for real wind fields, which are typically neither stationary nor uniform. One could attempt to resolve these effects by intercomparing the models for a few selected real wind cases for which adequate wind and wave field observations exist. However, a detailed interpretation of model performance in such cases will generally be rather difficult because of the simultaneous superposition of a variety of physical and geometrical influences (e.g., nonstationarity, inhomogeneity, turning winds). A more promising strategy for revealing the basic differences between models is to run the models instead for a series of hypothetical, idealized wind fields defined specifically to expose the models separately to particular critical influences, such as asymmetrical boundary conditions, suddenly changing wind directions, or discontinuous spatial wind distributions. This approach was adopted for all test cases, with the exception of the problem considered in Case VI.

As the emphasis was on the physics of the models rather than operational applications, all case studies were carried out in cartesian geometry and for an infinite-depth ocean.

In assessing the results of the case studies it should be kept in mind that the cases were specifically designed to emphasize critical aspects of the models. The models may therefore be expected to exhibit less divergence and perform significantly better, on the average, for real wind fields than for the cases studied here.

It is a pleasure to acknowledge the support of the many national funding agencies which made this project possible. Particular thanks are due to the United States Office of Naval Research for providing important travel and logistic support for this project through Grant N00014-80-C-0440, and to the Ministry of Transport and Public Works–Rijkswaterstaat, The Netherlands, for redrawing the greater part of the figures.

Chapter 1

Introduction: The Spectral Energy Balance

The evolution of a surface wave field in space and time is governed by the basic transport or energy balance equation

$$\frac{\partial F}{\partial t} + \mathbf{v} \cdot \nabla F = S \equiv S_{\text{in}} + S_{\text{nl}} + S_{\text{ds}} \tag{1.1}$$

where $F(f, \theta; \mathbf{x}, t)$ is the two-dimensional (2d) wave (surface variance) spectrum, dependent on frequency f and propagation direction θ; $\mathbf{v} = \mathbf{v}(f, \theta)$ is the (deep-water) group velocity; and the net source function S is represented as the sum of the input S_{in} by the wind, the nonlinear transfer S_{nl} by resonant wave–wave interactions, and the dissipation S_{ds}. The refractive terms representing the interaction with slowly variable currents, which are normally small, are not included in the propagation operator.

All models considered in this study compute the full 2d wave spectrum by numerical integration of (1.1), or some approximation to (1.1). The models differ primarily in the form assumed for the source function S. However, the structure of S can influence the choice of representation of F and thereby, indirectly, the numerical method used to integrate the transport equation.

The first attempt to develop a wave prediction model based on the differential transport equation (1.1), rather than on empirical integral relations between a few characteristic wave and wind field parameters, was made by Gelci *et al.* (1957). At the time the model was developed, very little was known about any of the three source function components. However, the theories of wave generation proposed in the same year by Phillips (1957) and Miles (1957)—together with Phillips' (1958) concept of a universal equilibrium range of the spectrum, and Hasselmann's (1960, 1961, 1962, 1963a, b) analysis of the nonlinear energy transfer due to resonant wave–wave interactions—then provided a broader theoretical framework for the wave models which were developed in the 1960s and further modified in the 1970s (Pierson *et al.*, 1966; Barnett, 1968; Gelci and Devillaz, 1970; Ewing, 1971; Gelci and Chavy, 1978).

The input source function in these models was normally represented as a superposition

$$S_{\text{in}} = A + B \cdot F \tag{1.2}$$

of a term A representing Phillips' external turbulent pressure forcing and a term $B \cdot F$ corresponding to Miles' [or Jeffreys' (1924)] linear feedback mechanism.

The dissipation source function S_{ds} was generally treated in the windsea region of the spectrum as an on–off limiter, which prevented the spectrum from exceeding a prescribed universal saturation level and became effective only when the spectrum approached this level.

The nonlinear transfer S_{nl} was either neglected entirely or (Barnett, 1968; Ewing, 1971) was incorporated as a relatively simple parameterization of Hasselmann's (1963b) early computations of the nonlinear transfer for a fully developed Neumann spectrum. Although the nonlinear transfer in Barnett's and Ewing's models could become comparable with the wind input terms under certain circumstances, its qualitative effect was generally relatively minor, as the nonlinear growth enhancement tended to occur in the same region of the spectrum (at the peak) as the input source function.

These first-generation wave models were applied successfully for many years, and are largely still in operation today. However, doubts had existed since their inception as to whether they really represented the physics of wave generation correctly. To reproduce the observed wave growth, Phillips' A term in (1.2) had to be chosen to be several orders of magnitude greater than estimates of the term based on turbulent pressure measurements. The factor B in the second term similarly had to be increased almost an order of magnitude beyond the theoretical estimates of Miles or Jeffreys. The models were further unable to explain the pronounced overshoot phenomenon of a growing windsea first observed by Barnett and Wilkerson (1967) and subsequently confirmed by many other workers.

These questions were resolved in the early 1970s through extensive field measurements of wave growth under carefully selected, uniform fetch-limited wind conditions (Mitsuyasu *et al.*, 1971; Hasselmann *et al.*, 1973). The

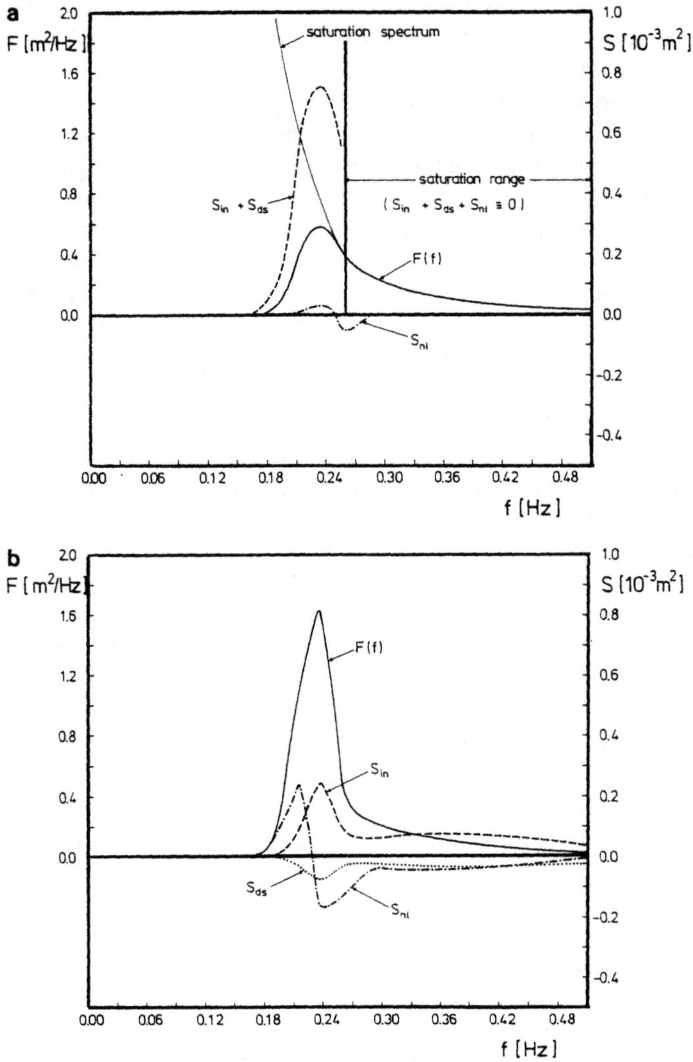

Fig. 1.1. Energy balance for first-generation (a) and second-generation (b) models. The source functions shown in panel b were computed with the model EXACT-NL (cf. Table 1.1). S_{in}/F is nearly an order of magnitude greater in panel a than in panel b, while S_{nl}/F is an order of magnitude smaller and has a different distribution as a consequence of the flatter spectral peak.

analysis of these data, corroborated by wind–wave tank experiments (Mitsuyasu, 1968a, b, 1969), led to a completely changed view of the spectral energy balance of a growing windsea, as indicated in Fig. 1.1 (cf. also the review by Barnett and Kenyon, 1975).

According to the revised picture, the principal source of energy during the main growth phase of the waves on the low-frequency forward face of the spectrum was the nonlinear energy transfer from higher-frequency components to the right of the spectral peak, rather than direct wind forcing. The estimated residual wind input on the forward face of the spectrum was nearly an order of magnitude smaller than that assumed in the first-generation models, and was therefore consistent both with earlier theoretical estimates and with later direct measurements of the work done by the surface pressure fluctuations on the moving wave surface by Snyder *et al.* (1981).

The spectral peak of a growing windsea was observed to be enhanced by a factor of approximately three relative to the form for a fully developed spectrum. This explained the unexpectedly high nonlinear transfer rate, which was computed to be an order of magnitude greater than that originally evaluated by Hasselmann (1963b) for a fully developed spectrum. Apart from being responsible for the rapid growth rates on the forward face of the peak, the nonlinear transfer was also found to control the shape of the spectrum, including the development of the peak itself. The overshoot phenomenon was explained naturally by the existence of this very sharp peak, significantly more pronounced than for a fully developed spectrum, which migrated toward lower frequencies as the wave spectrum developed.

At higher frequencies, a universal equilibrium range was not confirmed. Although the frequency spectrum approximately followed Phillips' f^{-5} power law, the level of the spectrum (Phillips' constant α) decreased with increasing fetch. [Kawai *et al.* (1977) have since proposed that windsea growth data may be represented with less scatter in terms of an f^{-4} power law, as proposed by Toba (1973) with a fetch-independent constant $\tilde{\alpha}$ proportional to $u_* g$, where $u_* = (\text{kinematic wind stress})^{1/2}$ represents the friction velocity (cf. also Kahma, 1981).]

The changed interpretation of the spectral energy balance required a basic restructuring of wave models. Whereas the first-generation wave models had been based on primarily decoupled wave dynamics, the second-generation wave models now had to cope with a dominantly nonlinear, coupled set of wave propagation equations. The following chapter summarizes the principal differences between first- and second-generation models, and the various approaches pursued in the treatment of the nonlinear coupling in second-generation models.

Chapter 2

Model Classes

In the first-generation wave models, each spectral component evolves essentially independently of all other components in accordance with the linear input source function (1.2) until it approaches its limiting saturation level, which is again defined independently of the energy in other spectral components by a universal equilibrium distribution. The weak coupling by the nonlinear transfer, if considered at all, represents a relatively small modification and is parameterized rather simply in terms of only one or two integral spectral parameters. It is therefore natural to represent the spectrum in these models directly as a two-dimensional discretized array of frequency–direction energy "packets," each of which propagates at its appropriate group velocity along its own ray path, responding only to the wind history sampled along this particular path. The numerical integration is carried out either in natural characteristic coordinates along individual rays, or in terms of a discretized advection operator within a common grid point representation for all wave components. We refer to these essentially linear, decoupled first-generation wave models as *decoupled propagation* (*DP*) *models*. [Although Barnett (1968) and Ewing (1971) include the nonlinear transfer explicitly and therefore have to contend with some of the numerical complications of coupled rays encountered in the later second-generation models, we nevertheless class these models physically as first- rather than second-generation models since the wave growth and spectral form are dominated by the wind input rather than the nonlinear transfer.]

In the revised energy balance of the second-generation wave models, the independent evolution of individual wave components is effectively prevented by the coupling through the nonlinear energy transfer. For a normal, relatively slowly varying wind field, the redistribution of energy by the nonlinear transfer is sufficiently rapid relative to advection and the other source functions that a universal, quasi-equilibrium spectral distribution of the JONSWAP form (cf. Hasselmann *et al.*, 1973, or Chapter 20) is established whose shape is largely independent of these other processes and which can be approximately characterized by a single, slowly changing scale parameter, such as the total energy E or peak frequency f_p (Hasselmann *et al.*, 1976).

However, for strongly nonuniform wind fields (e.g., fronts, intense tropical cyclones, or midlatitude storms—and most of the case studies considered in this volume), the advective terms are able to compete with the nonlinear energy redistribution, and significant deviations from the normal quasi-equilibrium spectral distribution can occur. The proper description of the wave field evolution in these cases requires not only an adequate representation of the wind input and dissipation source functions, S_{in} and S_{ds}, but above all a reliable parameterization of the nonlinear source function S_{nl} for arbitrary spectral distributions.

Exact computations of the three-dimensional nonlinear Boltzmann integral expression for S_{nl} (Hasselmann, 1962) are unfortunately still too time-consuming for incorporation into a general numerical wave model. (However, examples of transport equation integrations with exact calculations of S_{nl} are given in Cases II and VII for three particularly simple wind fields involving only one integration coordinate.) Some form of approximation to S_{nl} is therefore needed. Ideally, this should be in the form of some relatively simple nonlinear operator which can be applied to an arbitrary 2d spectral distribution $F(f,\theta)$ to yield an approximate source function $S_{nl}(f,\theta)$ containing the same number of degrees of freedom as used to describe the spectrum. Although nonlinear operators with this property are currently being investigated, no high-resolution operator approximation to S_{nl} was available for the models considered in this test.

The present parameterizations of S_{nl} contain only a few degrees of freedom. They are normally obtained by fitting some suitably chosen family of functions, dependent on a small number of adjustable parameters $c_1, c_2, \ldots, c_n$, to sets of exact nonlinear transfer computations made for a similarly restricted family of spectral distributions, which can be characterized by a few parameters $a_1, a_2, \ldots$ (Barnett, 1968; Mitsuyasu, 1968b; S. Hasselmann *et al.*, 1985). To estimate the nonlinear transfer for a given spectrum, the relevant spectral parameters $a_1, a_2, \ldots$ (for example, the total energy E, peak frequency f_p, mean direction θ, peak enhancement factor γ) are first evaluated. The appropriate parametrical expression for the set of parameters $c_1, c_2, \ldots, c_n$ corresponding to the set of parameters $a_1, a_2, \ldots$ is then substituted in place of the exact expression for S_{nl}.

Since the evolution of the windsea spectrum is strongly controlled by the nonlinear transfer, it can be argued that there is little advantage in representing the spectrum by more degrees of freedom than are available to parameterize the nonlinear transfer. In the so-called *parametrical windsea models*, the growth of the windsea part of the spectrum is accordingly expressed in terms of a small set of coupled transport equations describing the evolution of only a few characteristic spectral parameters. A general method for projecting the transport equation in the complete (f,θ) representation onto an approximate parameter space representation is described in Hasselmann *et al.* (1976).

For many applications, an acceptable first-order model may be developed in the form of a simple one-parameter prognostic transport equation describing the slow evolution of the single scale parameter characterizing the quasi-equilibrium windsea distribution for slowly varying wind fields (Hasselmann *et al.*, 1976; Sanders, 1976; Toba, 1978). Higher-order corrections can be systematically incorporated into such models by introducing further spectral parameters. In the limit of a large number of parameters one ultimately recovers the same detail of representation as with a traditional discrete spectral model. However, existing parametrical windsea models contain far fewer free parameters (normally one, maximally six in the present study) than normal discrete spectral models (typically several hundred frequency–direction energy bins).

A parametrical description of the wave field is appropriate only for the windsea region of the spectrum in which the nonlinear energy transfer is dominant. Normally, a wave spectrum consists of a superposition of a windsea spectrum and lower-frequency swell components which are no longer being actively generated by the wind. The swell energy level is generally sufficiently low that the nonlinear coupling between swell components or between swell and windsea is negligible (cf. Hasselmann, 1963b). The evolution of swell is controlled primarily by advection, perhaps with some relatively weak linear damping, and can therefore best be represented in the traditional framework of a decoupled discrete spectral model. Parametrical windsea models are therefore normally combined as hybrid models with standard discrete spectral representations for the swell components. We shall refer to these models in the following discussion as *coupled hybrid (CH) models*.

CH models may be expected to encounter difficulties in the windsea–swell transition regime of the spectrum in which the nonlinear energy redistribution is neither negligible nor dominant. Such transition regimes arise whenever the wind speed decreases or the wind direction turns.

In view of this difficulty (among other reasons) some modelers prefer to forego the basic informational economy offered by the parametrical approach and retain the traditional discrete spectral representation for the

Table 2.1
Models Used in Intercomparison Study

Model abbreviation (chapter number)	Institution	Application	Model class	Spatial representation
MRI (15)	Meteorological Research Institute, Japanese Meteorological Agency, Tsukuba, Japan	Operational forecast, North Pacific; regional wave statistics	DP	Grid points
VENICE (16)	Istituto per lo Studio della Dinamica delle Grandi Masse, Venice, Italy Associazione Studio Correnti Onde, Padua, Italy	Regional wave statistics	DP	Characteristics to output points only
NOWAMO (17)	Norwegian Meteorological Institute, Bergen, Norway	Operational forecast, Northeast Atlantic	CH	Windsea, grid points; swell, characteristics to output points only
GONO (18)	Royal Netherlands Meteorological Institute (KNMI), De Bilt, The Netherlands	Operational forecast, North Sea	CH	Windsea, grid points; swell, characteristics to output points only
TOHOKU (19)	Geophysical Institute, Tohoku University, Sendai, Japan	Regional wave statistics	CH	Grid points

HYPA (20)	Institut für Meereskunde, Hamburg, Federal Republic of Germany	Regional wave statistics; (test operational forecast North Atlantic and North Sea)	CH	Windsea, grid points; swell, characteristics
BMO (21)	British Meteorological Office, Bracknell, United Kingdom	Operational forecast, Atlantic, North Sea, North Pacific, Mediterranean; regional wave statistics	CD	Grid points
SAIL (22)	Atlantic Oceanographic and Meteorological Laboratory, National Oceanic and Atmospheric Administration, Miami, Florida, USA	Regional wave statistics	CD	Grid points
DNS (23)	Scripps Institution of Oceanography, La Jolla, California, USA Naval Ocean Research and Development Activity, NSTL Station, Mississippi, USA	Regional wave statistics	CD	Characteristics
EXACT-NL (24)	Max Planck Institute of Meteorology, Hamburg, Federal Republic of Germany	Nonoperational, one-dimensional integration only	CD	One-dimensional grid points

entire spectrum, including both windsea and swell regions. We shall refer to models of this type as *coupled discrete (CD) models*. A uniform discrete spectral representation is clearly conceptually simpler and easier to implement than a hybrid model. However, it must be kept in mind that the potential advantage of a more flexible representation of the windsea spectrum and a uniform representation of the swell–windsea transition regime cannot be properly exploited as long as the parametrical representation of the nonlinear energy transfer remains limited, as in present CD models, to relatively few degrees of freedom.

The mismatch between the number of degrees of freedom used in the description of the spectrum and in the parameterization of the nonlinear transfer gives rise to another problem. In the true energy balance, local perturbations in the spectrum tend to become smoothed out by the nonlinear transfer. However, parameterizations of the nonlinear transfer limited to only a few degrees of freedom are generally unable to damp out all forms of perturbations permitted by the discrete representation of the spectrum. A straightforward combination of such parameterizations of S_{nl} with reasonable analytical expressions for the wind input S_{in} (e.g., from Snyder *et al.*, 1981) and dissipation function S_{ds} (e.g., in the form of a turbulent viscosity) usually results in the unstable development of unrealistic spectral distributions. In the CD spectral models considered in this study the instability problem is resolved by artificially restraining the spectrum to accord with a predetermined spectral distribution in the unstable regions of the spectrum (normally for frequencies beyond the spectral peak). However, this effectively reduces a CD model to a CH model.

In practice, the principal distinction between present CD and CH models lies in the division between the discrete and parametrical regions of the spectrum. In hybrid models, the entire windsea spectrum is treated parametrically and the discrete representation is limited to the swell components only. In discrete models, the discrete representation normally includes both the swell and windsea peak, only the tail end of the spectrum beyond the peak being treated parametrically. (For all models a parametrical description is ultimately required at very high frequencies for which the dynamical response time becomes comparable with or smaller than the discrete time step of the model. However, the parametrical regions of present discrete models begin at significantly lower frequencies than this numerical time resolution cut-off.)

Table 2.1 lists the various models of type DP, CD, or CH considered in this study. The two DP models MRI and VENICE may be regarded as typical representatives of this class. Similar operational models which could not be made available for this study include the FNOC model (Lazanoff and Stevenson, 1975), the DHI model (Sand *et al.*, 1982), and the DSA V model

(Gelci and Chavy, 1978). These models all represent pure DP models without any parameterization of the nonlinear transfer, in contrast to the weakly coupled first-generation models of Barnett (1968) and Ewing (1971) (also not considered in this study). Most of the second-generation models of type CD or CH currently in operation are included in the table. In addition to the set of tests carried out with the nine operational models listed, a few integrations of the transport equation were made with the model EXACT-NL based on exact computations of the nonlinear energy transfer (Cases II and VII). With this model the transport equation could be integrated explicitly without any form of parametrical restriction up to the numerical time resolution cut-off frequency.

An overview of the main features of the different models and a discussion of some of the basic differences in the prediction properties of the three classes of models is given in Chapters 4–6. An extended summary of the individual models can be found in the chapters in Part II, which provide further references to more detailed presentations.

Chapter 3

Test Strategy

3.1. QUESTIONS ADDRESSED BY THE TESTS [WITH SUMMARIZED ANSWERS]

Although all nine operational models considered in this study have been validated against similar data, inspection of the physics and parameterizations of individual models reveals substantial differences, even between models of the same class. In addition, the numerical implementation varies significantly from one model class to another. This raises the following questions:

- □ Does the independent validation of different models using significantly different physics indicate that wave prediction models are insensitive generally to the assumed physics?

  [We find that this is not the case. For example, it is impossible to tune a first-generation DP model to agree with a second-generation CH or CD model for both fetch-limited and duration-limited wave growth, as also pointed out by Resio and Vincent (1979).]

- □ What are the differences in practice between the alternative spectral representations adopted in the second-generation models of type CH (hybrid) and CD (discrete)?

  [In the present stage of development we find little real distinction between the two. To exploit fully the potential advantage of the more detailed, but more time-consuming, CD models, generalized parameterizations of the nonlinear transfer are needed.]

□ Can critical elements in the model physics be identified which are responsible for strong divergences between the predictions of different models?

[The case studies were designed to reveal these. Major sources of uncertainty were found to be associated with the response to changing wind directions and the modeling of the transition region between windsea and swell. However, surprisingly large discrepancies were also found in the basic fetch- and duration-limited growth curves for a uniform wind field.]

□ Having identified the most critical weaknesses of present wave models, what should be done to improve our understanding and modeling of the wave spectral energy balance?

[A number of recommendations are given in the individual case studies and are collected in Chapter 13. Suggestions are made for the analysis of selected sets of existing data, the collection of new observations, and the development of improved numerical parameterizations.]

3.2. TEST CASES

It was felt that it would not be possible to separate and resolve these questions satisfactorily by comparing the models against real data. Instead, a set of seven hypothetical test wind fields was designed to focus separately on various critical properties of the models. The cases were ordered with respect to increasing complexity (except for the later addition of Case VII, which will be discussed prior to Case V). All models were run on identical grids, except when this was not possible because of technical constraints (such as the limitations of individual computer facilities).

Case I (advection test) was a pure swell propagation experiment designed to provide a comparison of the numerical advection schemes of discrete spectral models. A discussion of these results is not given here, however, as it was decided in retrospect that, although all models yielded acceptable numerical approximations to the pure advection of infinitely narrow-band spectral components, none of the numerical forward advection schemes currently in use satisfied the spatial dispersion properties appropriate for the propagation of the finite-sized spectral bins of real, finite-resolution discrete spectral models. The dispersion properties of the different frequency and wavenumber components contained in a finite-sized spectral bin lead to a linear increase of the spatial extent of a wave packet which is located initially at a single grid point by an amount which is proportional to the distance propagated multiplied by the spectral bandwidth and divided by the group velocity. The problem is described in more detail in Appendix B, where a nonlocal propagation operator with the desired linear dispersion properties is derived. The difficulties can be avoided in models, such as the VENICE, GONO, and NOWAMO models, which use backward ray-tracing techniques to reconstruct the output field at a few selected locations.

We have also excluded from detailed consideration the linear attenuation of swell. Although the assumed swell damping rates vary widely from one model to another, the question is basically straightforward from a modeling viewpoint and can be addressed meaningfully only by comparison with data. Most of the test cases were relatively insensitive to differences in swell damping.

Case II (fetch- and duration-limited growth) concerned the growth of a wave field for a uniform, stationary wind blowing orthogonally off a straight shore. The sea state was initially zero. For large fetch the evolution of the wave field with time provided idealized duration-limited growth curves, while in the asymptotic stationary limit the dependence of the wave field on distance offshore yielded idealized fetch-limited growth relations. The results from this case provided a reference base for discussing the effects of the more complicated wind field geometries considered in the remaining case studies.

Case III (slanting fetch) represented a generalization of Case II to an offshore wind blowing at an angle of 45° to the coast. The experiment was designed primarily to test the directional response of the models for the simplest cast of a uniform wind field in which an asymmetry is introduced by the boundary condition.

Case IV (half-plane wind) was intended to test the propagation of swell away from the side of a laterally bounded wind field into a neighboring calm region. The case also provided information on the effect of a lateral wind field boundary on the generation of waves within the wind field region.

Case V (diagonal front) concerned the propagation of wind waves across a diagonal front where the wind turned suddenly by 90° from a parallel to a cross-wave direction. The purpose of the experiment was to test the directional response of the models to sudden changes in the wind direction. Because of the inhomogeneity of the wave field, however, the experiment actually represented a rather complex superposition of directional response and advection effects. To separate the two, Case VII was later added to the set of experiments.

Case VI (stationary and moving hurricane) represented the most complex wind field considered. It was included to test the performance of the models under extreme but nevertheless realistic wind conditions. Most of the critical elements of the models which were investigated separately in the other case studies came into play simultaneously in this example. The prior analysis of the model response under idealized wind geometries helped clarify the general behavior—and rather significant differences—of the models under these complex conditions.

Case VII (90° change in wind direction) represented a simpler analogue to Case V in which the advection effects were removed by considering a nonstationary rather than an inhomogeneous wind field. At a given stage in the development of a duration-limited windsea, the direction of a uniform wind field is suddenly turned by 90° into the cross-wave direction, remaining constant thereafter. Since the wave field remains homogeneous throughout, only one integration variable (the time) enters, rather than the two spatial coordinates of Case V.

A detailed discussion of the individual cases is given in Chapters 7–12. The principal conclusions and recommendations inferred from each case study are presented at the end of the appropriate sections and are collected again in the summary (Chapter 13).

3.3. NONDIMENSIONALIZATION

In all test cases except Case VI (stationary and moving hurricane) a nominal reference wind speed of $U_{10}(\equiv U) = 20$ m/s at 10 m height above the surface was assumed. The numerical value of U is in fact immaterial, since all model results can be expressed as dimensionless functions of the nondimensional space and time coordinates

$$\tilde{\mathbf{x}} = g\mathbf{x}/U^2, \qquad \tilde{t} = gt/U \tag{3.1}$$

or the equivalent variables

$$\mathbf{x}^* = g\mathbf{x}/u_*^2, \qquad t^* = gt/u_* \tag{3.2}$$

where

$$u_* = \tau^{1/2} = c_d^{1/2} \cdot U \tag{3.3}$$

represents the friction velocity corresponding to the (kinematic) surface wind stress $\tau = c_d U^2$.

We have chosen the variables $(\mathbf{x}^*, t^*)$ rather than $(\tilde{\mathbf{x}}, \tilde{t})$. The transformation between the two sets of variables depends on the drag coefficient c_d. Despite extensive investigations for many years, disagreement still exists regarding the dependence of c_d on local atmospheric boundary layer and sea state parameters. Except for Case VI, we have assumed a drag coefficient $c_d = 1.83 \times 10^{-3}$ for all models (corresponding to the value used in the SAIL model for $U_{10} = 20$ m/s). In the hurricane test Case VI, each model applied its own particular drag coefficient relation dependent on the variable local wind speed.

The value of the drag coefficient is important for our intercomparison study, since some models express the input source function S_{in} in terms of the wind speed U at some reference level (usually 10 m or the Pierson–Moskowitz reference height, 19.5 m) while other models use the friction velocity. In either case, the relevant velocities normally have to be related in practice to the geostrophic velocity, which is provided as primary driving field by the atmospheric diagnostic or forecast model. We discuss these differences in more detail in the context of Case II (fetch- and duration-limited growth).

A similar choice between U and u_* for the nondimensionalization of the wave model variables must be made. We have nondimensionalized here also in terms of u_*, in the form

$$E^* = g^2 E u_*^{-4} \quad \text{(Total variance)} \tag{3.4}$$

$$f_p^* = u_* f_p g^{-1} \quad \text{(Peak frequency)} \tag{3.5}$$

$$F^*(f) = F(f) g^3 u_*^{-5} \quad \text{(One-dimensional spectrum)} \tag{3.6}$$

$$F^*(f,\theta) = F(f,\theta) g^3 u_*^{-5} \quad \text{(Two-dimensional spectrum)} \tag{3.7}$$

In some cases it is convenient to normalize the wave model response with respect to variables defined through the fully developed Pierson–Moskowitz (1964) spectrum:

$$F_{PM}(f) = \alpha g^2 (2\pi)^{-4} f^{-5} \exp\left[-5/4 (f_{PM}/f)^4\right] \tag{3.8}$$

where $\alpha = 0.0081$ and the Pierson–Moskowitz peak frequency (referred to in the following more briefly as the "wind frequency") is given by

$$f_{\mathrm{PM}} = 0.140 g/U_{19.5} \approx 0.13 g/U_{10} \tag{3.9}$$

The total variance of the PM spectrum is

$$E_{\mathrm{PM}} = \alpha g^2 (2\pi f_{\mathrm{PM}})^{-4}/5 \approx 3.64 \times 10^{-3} U_{10}^4 g^{-2} \tag{3.10}$$

For the drag coefficient $c_{\mathrm{d}} = 1.83 \times 10^{-3}$, equations (3.8)–(3.10) yield

$$f_{\mathrm{PM}}^* = 5.56 \times 10^{-3} \tag{3.11}$$

$$E_{\mathrm{PM}}^* = 1.09 \times 10^3 \tag{3.12}$$

The nondimensional maximal spectral density of the one-dimensional spectrum is given by

$$F_{\mathrm{PM}}^*(f_{\mathrm{PM}}) = 2.80 \times 10^5 \tag{3.13}$$

Chapter 4

Decoupled Propagation Models

The principal characteristics of the two DP models considered in this study (VENICE and MRI) are listed in Table 4.1. Details and further references can be found in the model descriptions given in Part II. For both models, wave growth is strictly decoupled up to the saturation limit.

A basic property of strictly decoupled models [which applies also to first order for the weakly coupled first-generation models of Barnett (1968) and Ewing (1971)] is that the differential time and space scales, dt and ds respectively, describing the evolution of any given spectral component along its ray path, are related through $ds = v\,dt$, where the group velocity v is a constant for a given wave group. It follows that the laws for fetch-limited wave growth under uniform, stationary wind conditions can be immediately translated into the corresponding duration-limited growth laws simply by replacing the fetch variable x for each wave component by vt (cf. Case II).

Thus, if we characterize the stage of development of a fetch-limited spectrum at fetch x, for example, by the frequency $f_{1/2}$ of the spectral component for which the spectral density has just reached half its fully developed value (this corresponds to a point about half way up the forward face of the spectrum), the equivalent duration t needed to reach the same stage of development for a duration-limited spectrum is given by

$$x(f_{1/2}) = v(f_{1/2})t(f_{1/2}) \tag{4.1}$$

The fetch-limited spectrum at $x(f_{1/2})$ will not be exactly identical in shape to the equivalent duration-limited spectrum at $t(f_{1/2})$ because of the frequency dependence, through the group velocity, of the ratio of the growth rates with respect to fetch and duration. However, the spectra at equivalent

Table 4.1
Source Functions S of Decoupled Propagation Models[a]

Model	Growth stage $F \le F_\infty$, $\lvert\theta_w\rvert \le 90°$	Swell decay, propagation in wind direction $F > F_\infty$, $\lvert\theta_w\rvert \le 90°$	Swell decay, opposing wind $90° < \lvert\theta_w\rvert < 180°$
MRI	$S = (A + B \cdot F)[1-(F/F_\infty)^2]S_2$	$S = -Df^4F$	$S = -(B' + Df^4)F$
	where $S_2 = \cos^2\theta_w$ spreading factor and $A = A(f, U)$, $B = B(f, u_*)$	$D = 1/3600\ \sec^3$	$B' = B(\theta + 180°)$
VENICE	$S = (A + B \cdot F)$	$S = -(F - F_\infty)\gamma$	$S = BF$
	$A = \dfrac{(0.9)\times 10^{-11} \cdot (U\cos\theta_w)^4}{g^2}$ $B = 6.2\times 10^{-3} f\left(\dfrac{U\cos\theta_w}{c} - 0.9\right)$	$\gamma = \begin{cases} \dfrac{(U/c)^2}{3600}\sec^{-1} & \text{for } \dfrac{U}{c} > \dfrac{1}{1.3} \\ 0 & \text{for } \dfrac{U}{c} < \dfrac{1}{1.3} \end{cases}$	(note that $B < 0$ for $90° < \lvert\theta_w\rvert \le 180°$)

[a]Notation: U = wind velocity

θ_w = angle between wind and wave direction

c = phase velocity

F_∞ = fully developed spectrum = (Pierson–Moskowitz spectrum) $\begin{cases} (S_2(\theta)) & \text{for MRI model} \\ (S_4(\theta)) & \text{for VENICE model} \end{cases}$

$S_2(\theta) = \frac{2}{\pi}\cos^2\theta$ for $\lvert\theta\rvert < \frac{\pi}{2}$, zero otherwise ("$\cos^2$-spreading function")

$S_4(\theta) = \frac{8}{3\pi}\cos^4\theta$ for $\lvert\theta\rvert < \frac{\pi}{2}$, zero otherwise ("$\cos^4$-spreading function")

fetch- and duration-limited growth stages nevertheless exhibit the same general shape and, by our definition of $x(f_{1/2})$ and $t(f_{1/2})$, coincide in the position of the forward face (as determined by the half-energy point).

Equation (4.1) may be compared with the equivalent fetch–duration relation for coupled hybrid models [cf. Chapter 5, equation (5.7)]. In this case, the group velocity $v(f_{1/2})$ is replaced by the smaller effective advection velocity $\frac{8}{9}\bar{v}$, where $\bar{v}$ represents a mean energy propagation velocity averaged over the spectrum, and the factor $\frac{8}{9}$ arises from averaging over the past history of the developing windsea. Similar relations hold for coupled discrete models. The discrepancy between (4.1) and (5.7) implies that it is not possible to tune first- and second-generation wave models to yield the same growth laws for both fetch- and duration-limited wave growth, as has been pointed out also by Resio and Vincent (1979).

The independent evolution of different wave components along different ray paths results in another distinguishing feature of decoupled propagation models: the spectrum generally develops more frequency and directional fine structure than in a coupled model, in which the energy is being continually redistributed and smoothed by the nonlinear transfer. These differences are apparent in most of the case studies.

An advantage of DP models is that they can be applied very economically to predict the output wave field at only a few selected stations and times of interest. Because of the decoupling, there is no need to compute the energy of the great majority of wave components which do not propagate toward the selected stations or do not arrive at the station at the defined output times. The VENICE model (and the swell sections of the hybrid models GONO and NOWAMO, cf. Chapter 5) operate on this principal. The ray approach also lends itself readily to an extension to shallow water through the inclusion of refraction (as implemented, for example, in the VENICE model).

Chapter 5

Coupled Hybrid Models

5.1. THE WINDSEA SPECTRUM

Empirically, a parametrical description of the windsea spectrum could be motivated by the observation that the spectra of growing windseas appear to have approximately the same shape for a wide variety of generation conditions. The spectral distributions differ only with respect to the frequency and energy scales. Furthermore, the two scale parameters are found to be approximately interrelated when nondimensionalized in terms of g and the local wind speed U or friction velocity u_*. Thus the spectrum of a growing windsea can be characterized to first order by a single nondimensional variable, for example the nondimensional energy $E^* = Eg^2u_*^{-4}$. This determines all other nondimensional variables, such as the nondimensional peak frequency $f_p^* = u_* f_p/g$ and Phillips' "constant" α. The ratio of E^* to its value E_∞^* for a fully developed sea provides a unique measure of the "age" of the windsea.

The existence of a quasi-universal spectral shape for windseas has long been known, or at least assumed. It is implicit in the early empirical wave prediction formulas in use prior to wave prediction methods based on the spectral transport equation. It also provided the empirical basis for the first hybrid model by Haug (1968) (the NOWAMO model of this study, cf. Table 2.1), which was developed prior to our present understanding of the spectral energy balance. More detailed empirical investigations of the self-similarity hypothesis were made by Wilson (1965), Mitsuyasu *et al.* (1971, 1980), Toba (1972, 1978), Hasselmann *et al.* (1976), Kawai *et al.* (1977), Sanders and Bruinsma (1982), and other workers. A dynamical explanation of the quasi-self-similarity in terms of the shape-stabilizing properties of the nonlinear transfer was given by Hasselmann *et al.* (1973, 1976). The phenomenon is demonstrated explicitly by the integrations with the EXACT-NL model presented in this volume.

The evolution equation for the total energy E of the windsea spectrum can be obtained by integrating the transport equation over all frequencies and directions:

$$\frac{\partial E}{\partial t} + \nabla \cdot (\bar{\mathbf{v}} E) = S_E \tag{5.1}$$

where

$$\bar{\mathbf{v}} = \int \mathbf{v} F(f, \theta)\, df\, d\theta / E \tag{5.2}$$

is the effective mean propagation velocity for the total energy and

$$S_E = \int S(f, \theta)\, df\, d\theta \tag{5.3}$$

represents the projection of the net source function S onto the parameter E.

The energy advection velocity $\bar{\mathbf{v}}$ is uniquely determined as a function of E by the prescribed spectral shape and the empirical dependence of the second spectral scale parameter, f_p, say, on E. [Toba (1973), for example, proposes the empirical relation $E^* \sim f_p^{*-3}$, which is used in the TOHOKU model, cf. Table 5.1.] To close the equation, S_E must be prescribed as a function of E and the local wind U or friction velocity u_*. This is normally determined empirically.

Most of the hybrid models in use today are based on a one-parameter representation of the windsea spectrum and use the simple transport equation (5.1), or some form equivalent to (5.1). Of the four CH models considered in the present study, only the HYPA model uses more than one prognostic parameter, namely the five JONSWAP parameters and the mean wave direction $\bar{\theta}$. (The GONO model stores a second parameter representing "old windsea" which under certain conditions can influence the development of the real windsea. However, formally this effect should be classed as a swell–windsea interaction, which will be discussed later.)

A general technique for generalizing equation (5.1) by deriving from the full spectral transport equation a set of prognostic transport equations for an arbitrary number of free parameters is described in Hasselmann *et al.* (1976).

Table 5.1 lists the prognostic variables and the principal properties of the spectral shapes, scale relations, and source functions for the windsea sections of the hybrid models of this study.

Table 5.1
Windsea Description and Dynamics of Coupled Hybrid Models

Model	Prognostic variables	Spectral form: Frequency	Spectral form: Direction	Source function
NOWAMO	E	Neumann spectrum, f^{-4} beyond peak	$\cos^2\theta$	S_E determined from empirical growth curve. Some loss of windsea energy to swell occurs for turning winds.
GONO	E	Linear from minimum frequency to peak frequency, f^{-5} beyond peak	$\cos^2\theta$	S_E determined from empirical growth curve. Some loss of windsea energy to swell occurs for turning winds.
TOHOKU	$\varepsilon \equiv E^* = g^2 E/u_*^4$	f^4 forward face, f^{-4} beyond peak f_p $E^* = B(2\pi f_p^*)^{-3}$, $B = 5.1\times10^{-2}$	$\cos^2\theta$	$\frac{dE^*}{dt^*} = S_\varepsilon = \frac{3}{2}A\cdot\varepsilon^{1/3}[1-\mathrm{erf}(b\varepsilon^{1/3})]$ $A = 2.4\times10^{-4}$ $b = 0.12$
HYPA	JONSWAP parameters $a_1,\dots a_5 = f_m$, α, γ, σ_a, σ_b and $a_6 =$ mean wave direction $\bar{\theta}$	JONSWAP spectrum	$\cos^2\theta$	$S_2,\dots,S_5$ produce rapid shape stabilization; S_1 describes slower change of spectral scale parameter in accordance with one-parameter models. $S_6 \equiv S_{\bar{\theta}}$ produces exponential directional relaxation for small step-function changes in wind direction.

The one-parameter models GONO, NOWAMO, and TOHOKU yield qualitatively similar windsea predictions. The six-parameter HYPA model also yields similar results for slowly varying wind fields, but is able to develop more varied spectral shapes and directional lag effects for nonuniform wind fields. Of the additional prognostic parameters, the directional parameter $\bar{\theta}$ appears to be the most important (in the three one-parameter models, the mean wave direction is treated as a diagnostic variable which is not governed by a prognostic transport equation, but is assumed to adjust instantaneously—with some accompanying loss by radiation into swell—to the local wind direction).

As remarked in Chapter 4 with reference to the growth properties of decoupled propagation models, one of the distinguishing features between coupled and decoupled propagation models is the relation between fetch- and duration-limited wave growth. The corresponding relation for coupled propagation models may be illustrated by the TOHOKU model. Other coupled propagation models yield qualitatively similar relations (cf. Case II).

For the initial growth stage, the error function responsible for the transition to the fully developed state in the source function S_ε of the TOHOKU model (cf. Table 5.1) can be neglected, and one obtains the simple fetch- and duration-limited growth relations

$$E^* = \frac{9A}{4C}B^{1/3}\cdot x^* \qquad \text{(fetch- limited)} \tag{5.4}$$

$$E^* = (At^*)^{3/2} \qquad \text{(duration-limited)} \tag{5.5}$$

in which the constants A and B are given in Table 5.1 and the constant $C = \bar{v}/v_{\mathrm{p}}$, where v_{p} denotes the group velocity at the peak frequency f_{p}. For the TOHOKU spectral form, one finds $C = 5/2\pi = 0.796$. For the same stage of wave development (i.e., the same dimensionless energy E^*) the dimensionless fetch and duration are accordingly related by

$$x^* = \frac{4A^{1/2}B^{-1/3}}{9}\cdot Ct^{*3/2} \tag{5.6}$$

or, invoking the relation between the peak frequency and energy (Table 5.1) and equation (5.5),

$$x = \frac{8}{9}\bar{v}t \tag{5.7}$$

To compare (5.7) with the corresponding relation (4.1) for DP models, we note that for a given stage of spectral development the group velocity $v(f_{1/2})$ in (4.1) (where $f_{1/2}$ corresponds to a frequency near the middle of the forward face of the spectrum) is greater than the group velocity v_p at the peak, which in turn is greater than $\bar{v}$. Combining these inequalities with the factor 8/9 in (5.7), the ratio of fetch to duration for a given stage of development is seen to be greater for decoupled propagation models than for coupled hybrid or discrete models by a factor of the order of 1.5.

5.2. THE TRANSITION BETWEEN WINDSEA AND SWELL

If the local wind decreases in strength or turns in direction, part of the low-frequency region of a windsea spectrum can be transformed to swell. The reverse transition from swell to windsea occurs if the component of the wind blowing in the direction of a swell increases to a value greater than the swell phase speed (swell frequency f greater than the wind frequency). In practice, the transition between a strongly coupled windsea spectrum and a decoupled swell spectrum is a continuous process, even when the wind change is abrupt. Although the basic nonlinear coupling mechanism controlling this transition is understood theoretically, and acceptable expressions for the remaining source function for the wind input and dissipation have been proposed (or could be constructed), no operational wave model is yet able to describe the windsea–swell transition process satisfactorily. The main difficulty has been the construction of a sufficiently flexible parameterization of the nonlinear energy transfer.

In hybrid models, the simulation of the transition regime is necessarily strongly simplified, since the windsea spectrum is assumed to be strongly coupled, while the swell components are treated as completely decoupled. The transition is normally modeled as a simple energy interchange between the windsea and swell regions of the spectrum, accompanied by an instantaneous redistribution of energy within the modified windsea spectrum. Details of the transition criteria used in the four CH models considered in this study are given in the model descriptions in Part II and in the references listed therein. However, a few characteristic features may be mentioned here which are important for the interpretation of the case study results.

For turning winds, all CH models lose a portion of the windsea energy traveling at large angles to the new wind direction into swell. In the GONO, NOWAMO, and HYPA models, the energy loss is a continuous function of the rate of change of wind direction, whereas in the TOHOKU model an energy loss occurs only if the wind direction change is greater than 30° within a time step. Thus the TOHOKU model generally loses considerably less windsea energy in a turning wind than the other CH models. (This is also a characteristic feature of the CD model BMO, which has a similar constraint on the radiation of swell in slowly turning winds, cf. Chapters 10 and 11.)

The direction of the residual windsea is taken as the new wind direction in the one-parameter models GONO, NOWAMO, and TOHOKU, but is computed independently as the direction of the mean momentum of the residual windsea components in the multi-parameter HYPA model. The prognostic equation for the mean windsea direction $\bar{\theta}$ then tends to relax $\bar{\theta}$ back to the wind direction θ_w. For a continuously turning wind, the persistent lag between $\bar{\theta}$ and θ_w makes it easier for the HYPA model to split off windsea energy into swell at each time step than for the nonlagged one-parameter CH models.

For a decreasing wind, the lowest frequency components of the windsea spectrum are transferred to swell in all models when the windsea components travel faster than the wind. In one-parameter models this can also be expressed as the condition that the windsea energy exceeds the appropriate value for a fully developed windsea. Conversely, for increasing winds, a swell component is generally reabsorbed in the growing windsea when the windsea peak frequency becomes approximately equal to the swell frequency and the propagation directions of windsea and swell lie within a given narrow sector. In effect, the transition criteria windsea–swell and swell–windsea are then roughly symmetrical. The HYPA model represents an exception in the sense that the swell absorption condition depends only on the relative frequencies of the windsea and swell, independent of the propagation directions. An isotropic swell–windsea transition criterion is suggested by exact computations of nonlinear windsea–swell interactions, which were found to depend strongly on frequency but only weakly on direction (cf. Hasselmann, 1963b).

An interesting variant in the GONO and NOWAMO models is that the transfer of windsea energy into swell is not treated explicitly, but is reconstructed *a posteriori* in terms of "potential" swell components. At each grid point and time, the entire windsea spectrum is regarded as a potential source of swell radiating to the selected model output points. The real swell actually arriving at a given output point is taken, for a given frequency, direction, and arrival time, as the largest of all potential swell components in that spectral bin. (The NOWAMO model in fact retains only a single directional component for each frequency, namely the direction with the highest energy; this is then spread again into a $\cos^2\theta$ directional distribution.)

In terms of an equivalent explicit windsea–swell transfer algorithm, the GONO and NOWAMO schemes imply that swell, once generated, is able to propagate without loss or interaction through a local windsea provided it contains more energy than the corresponding component of the windsea spectrum. On the other hand, the swell energy is completely destroyed (without a corresponding increase in the windsea energy) if the energy of the windsea component exceeds the swell energy. Qualitatively, this is not very different from the explicit treatment of swell in the TOHOKU and HYPA models, except that the reabsorption of swell in the windsea is nonconservative (and the HYPA reabsorption criterion is isotropic).

A principal difficulty in the "potential" swell approach is that the influence of the swell on the windsea during the reabsorption process cannot be modeled. The GONO model partially remedies this problem by retaining an "old windsea" in the model (essentially a parameterization of newly created swell components in terms of a mean direction and energy) which can transfer energy to the windsea during reabsorption.

An advantage of the "potential" swell concept as implemented in the GONO and NOWAMO models is that, because the swell is not advected forward in time, the patchiness problem inherent in the normal numerical treatment of forward advection for discrete spectral models (cf. Appendix B) does not arise. The required mutual consistency in wavenumber and spatial resolution is achieved in these models by defining different sets of backward-tracing rays for the selected output points dependent on the distance from the source, the angular resolution of the rays increasing linearly with the distance.

5.3. SWELL PROPAGATION

All four CH models use characteristics to propagate swell. Assumptions about swell damping range from no damping (HYPA), to damping only in the presence of adverse wind (TOHOKU), to a general wind-independent damping (GONO and NOWAMO). Details are given in the individual model descriptions in the Part II. As pointed out in Chapter 3, however, the intercomparison study was not designed to investigate swell damping, and the test case results were accordingly relatively insensitive to the differences in the assumed swell attenuation rates.

Chapter 6

Coupled Discrete Models

All three CD models considered in this study—BMO, SAIL, and DNS—use a linear-feedback input source function $S_{in} = BF$, with reduced values of Miles' factor B relative to first-generation DP models, in accordance with the measurements of Snyder *et al.* (1981). (The BMO model also includes a Phillips forcing term, $S_{in} = A + BF$, but A is small and is relevant only for triggering growth in an initially calm ocean.) All models also invoke some form of limiting saturation spectrum which depends on the stage of development of the windsea. The models differ primarily in the details of the parameterization of the nonlinear transfer in the forward face and peak regions of the windsea spectrum and in the swell–windsea transition regime. As pointed out in Chapter 2, the limitations of presently available parameterizations of the nonlinear transfer restrict the effective number of degrees of freedom of CD models in the representation of the windsea spectrum. Thus the techniques used to simulate these processes, although containing a few more degrees of freedom, do not differ qualitatively from the methods described for CH models. The principal features of the models are listed in Table 6.1.

The EXACT-NL model is also included in Table 6.1 for reference. This model uses an explicit representation for all three source functions and computes the nonlinear transfer integral expressions exactly. Because of its large computing-time requirements, however, the model can be integrated only for simple wave field geometries for which only one integration coordinate occurs. The instabilities associated with the mismatch between the spectral representation and nonlinear parameterizations in present CD models do not arise, and the model can therefore be operated out to its natural high-frequency cut-off (typically two to three times the peak

Table 6.1
Source Functions of Coupled Discrete Models

Model	Linear growth	Nonlinear transfer	Saturation spectrum
BMO	$S_{in} = A + BF$	Total net energy input computed from S_{in} and S_{ds}, where $S_{ds} \sim -f^2 E^{1/4} F$, then redistributed in accordance with JONSWAP form.	JONSWAP spectrum; saturation is included in nonlinear transfer parameterization.
SAIL	$S_{in} = BF$	Net source function inferred from JONSWAP growth curves is applied if greater than S_{in}.	Modified JONSWAP-type distribution. Saturation is accompanied by directional energy redistribution.
DNS	$S_{in} = BF$	EOF representation of exact S_{nl} computations as function of peak enhancement factor γ and directional spread $\Delta\theta$.	Pierson–Moskowitz spectrum with sea-state-dependent saturation level.
EXACT NL	$S_{in} = BF$	Exact computations.	Saturation spectrum similar to observed spectrum generated by model with S_{in} and S_{nl} as given and a dissipation function $S_{ds} = -(7.8)\times 10^9 \left[\frac{f^4}{f_p^3}\right]\left[\frac{E(2\pi f_p)^4}{g^2}\right]^6$

frequency), for which the response time becomes comparable with the time step. The shape of the spectrum, including the saturation level, is not prescribed, but is determined by the internal dynamics of the model itself.

Of the operational CD models, BMO bears the strongest resemblance to a CH model and could almost be grouped in this class. At each growth step, the total energy input to the windsea is computed by integrating the sum of S_{in} and S_{ds} over the windsea spectrum. The net energy gain is then redistributed over the frequency and directional bands to yield a new spectral distribution in accordance with the JONSWAP spectrum appropriate to the computed nondimensional energy E^*. The directional distribution is also prescribed and is centered on the instantaneous wind direction. The model therefore operates in essentially the same way as a one-parameter CH model. A decaying wind successively decouples the low-frequency components of the windsea into swell, leaving a modified windsea spectrum with lower residual energy and an associated higher peak frequency. Radiation of windsea into swell also occurs for a turning wind, but only if there is a substantial change of wind direction in a time step.

The SAIL model also forces a growing windsea spectrum to conform closely to an empirical JONSWAP-type distribution. Here the net incremental growth at each time step is taken from the empirical fetch- limited spectral growth data and does not depend on an explicit representation of S_{nl}. For uniform wind fields the model automatically recovers the same fetch- and duration-limited growth curves as the CH models, which are similarly tuned. However, for turning winds the separate differential growth treatment of individual frequency-directional bands provides some memory for the directional distribution. Thus the mean direction of the windsea tends to lag the wind direction, enabling some windsea energy propagating at larger angles to the wind to radiate away into swell. The separate growth rates for different directional bands make this model more sensitive to lateral limitations of the wind field or asymmetric (slanting fetch) boundary conditions (Cases III and IV). In this respect the model exhibits characteristics more typical of DP models than CH models.

The Miles wind input term becomes effective in the SAIL model only in special situations in which it exceeds the empirical JONSWAP windsea growth rate. This can occur, for example, when an increasing wind regenerates former swell components whose frequencies lie above the local wind frequency but below the windsea peak frequency. (The CH model HYPA treats such components in the same manner within the swell section of the model.) A particular feature of the SAIL model is the treatment of the saturation region, which is not represented simply as a normal cut-off limit, but is associated with an angular energy redistribution. The saturation algorithm is also effective in reabsorbing swell components whose frequencies are smaller than or equal to the windsea peak frequency and which propagate in about the same direction as the windsea.

The DNS model is based on a source function consisting in the growth stage of a superposition of S_{in} and a parameterization of S_{nl} depending on two spectral shape parameters (the peak enhancement factor γ and the angular spread $\Delta\theta$). The parameterization was obtained from a principal component (EOF) analysis of an ensemble of exact computations of S_{nl} made for a series of spectra with different values of γ and $\Delta\theta$. To avoid instabilities, the net source function is applied only if it is positive. This in effect limits the prognostic range of the model to the forward face and peak of the spectrum. Beyond the peak frequency, the spectrum is limited by a Pierson–Moskowitz spectrum with variable saturation level depending on the stage of development of the windsea. As the DNS parameterization applies no constraints to the form of the growing windsea near the peak and on the forward face, it is able to simulate directional relaxation effects and the radiation of windsea into swell in the same way as the SAIL model. The reabsorption of swell into a windsea is based on a directionally dependent frequency criterion. The absorption process is nonconservative: wave components with frequencies greater than the windsea frequency are attenuated to conform to the limiting spectrum, which is centered on the wind direction. (This implies in particular that components with propagation directions at more than 90° to the wind direction are annihilated.)

In summary, the BMO model bears a close similarity, both in the physics and in the numerics, to one-parameter hybrid models, while SAIL and DNS exhibit characteristics of both CH and DP models. For turning winds or asymmetric wind field geometries the discrete representation of the SAIL and DNS models enables the models to develop more directional fine structure than one-parameter CH models or the HYPA model, in which the directional spreading distribution is prescribed relative to the prognostically determined mean windsea direction. However, it should be noted that the parameterizations of the nonlinear transfer in the CD models were not designed to simulate the energy redistribution in a skewed spectrum, and that the directional fine structures predicted by the SAIL and DNS models (and first-generation DP models), although probably correct qualitatively, cannot be relied upon in detail.

Chapter 7

Case II: Fetch- and Duration-Limited Growth

7.1. WIND FIELD, BOUNDARY CONDITIONS

The geometry of the wind field and boundary conditions are shown in Fig. 7.1. A stationary, homogeneous wind field with wind speed $U_{10} = 20$ m/s blows orthogonally offshore. The wind field is infinite in the lateral and downstream directions. The initial wave energy at time $t = 0$ is zero, and the spectrum at the coastline $y = 0$ remains zero for $t > 0$. The models were run until a stationary state was reached for the entire area.

7.2. PURPOSE OF THE TEST

Most models of the study were calibrated against stationary, uniform-wind, fetch-limited wave growth measurements. A comparison of the models under these conditions therefore provides the necessary baseline for the discussion of the more complex geometries of the other test cases. The test case also yields duration-limited wave growth curves (from the time evolution at large distances from the coast), which were used by some models (e.g., GONO) for their calibration.

Another purpose of the test was to investigate the relations between fetch- and duration-limited wave growth for different models. It was anticipated that these would reveal, for example, some fundamental differences between first- and second-generation models (cf. Chapters 3 and 4).

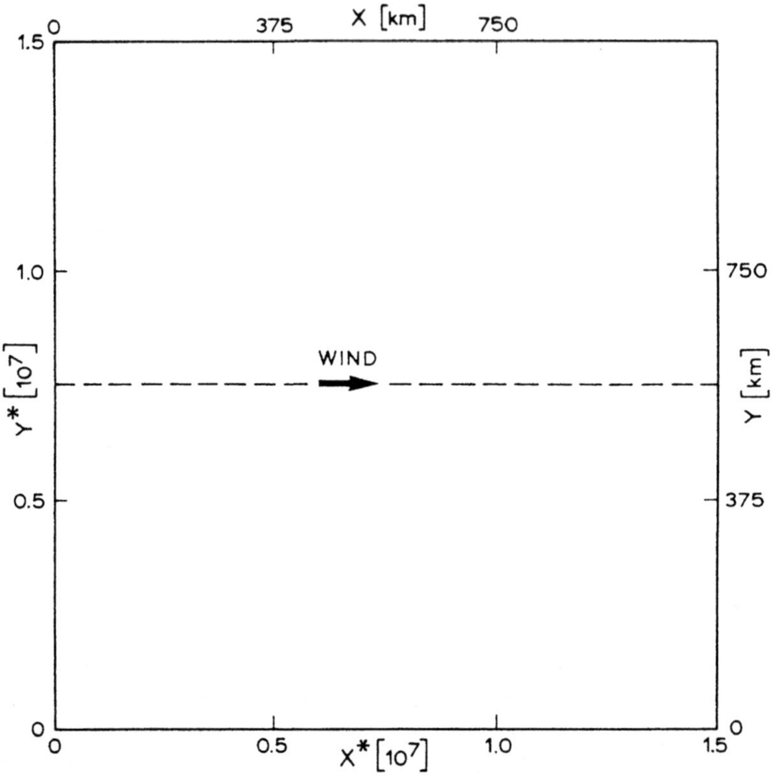

Fig. 7.1. Wind field geometry for Case II, fetch- and duration-limited growth study. Fetch-limited results were extracted along the dashed section, duration-limited development observed at (typically) $x^* = 1.3 \times 10^7$.

7.3. SPECTRAL SHAPE

The shapes of the one-dimensional windsea spectra generated by the models can be characterized in all cases by a sharp peak at some frequency f_p, a steep low-frequency forward face for $f < f_p$, and a more slowly decreasing high-frequency tail proportional to f^{-n}, where n is normally 5 or 4.

All models except TOHOKU and EXACT-NL assume equilibrium fully developed spectra for infinite fetch and duration (Fig. 7.2). In mose cases this is given by the Pierson–Moskowitz spectrum. However, it is not obvious that an exact equilibrium spectrum is ever achieved on ocean basin scales. In fact, the model EXACT-NL continues to evolve very slowly by transferring energy to still lower frequencies after f_p has decreased to the Pierson–Moskowitz peak frequency f_{PM}. The CH model TOHOKU exhibits similar behavior, but the wave growth is finally arrested at the stage of development shown in Fig. 7.2. The exceptional high-energy peak of the TOHOKU spectrum therefore simply reflects the fact that the peak frequency f_p for this limiting state is significantly lower than f_{PM}. The details of the transition to a quasi-fully developed state depend rather sensitively on the form of the input source function for waves whose phase velocity is close to or slightly greater than the wind speed. There exist few wave growth data to determine the asymptotic growth behavior at very large fetches, since the required very-large-scale uniform wind fields occur only rarely in regions in

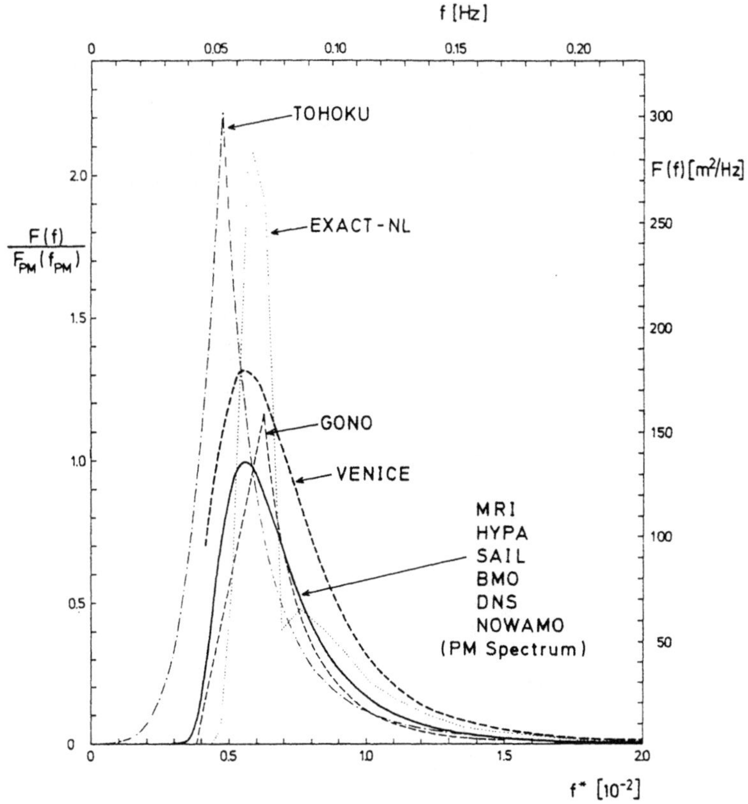

Fig. 7.2. Fully developed equilibrium spectra (Case II). EXACT-NL has no asymptotic equilibrium spectrum; the spectrum shown corresponds to a fetch of 1000 km. The TOHOKU spectrum also grows very slowly beyond the Pierson–Moskowitz spectrum, but wave growth is finally "artificially" arrested when the spectrum reaches the distribution shown.

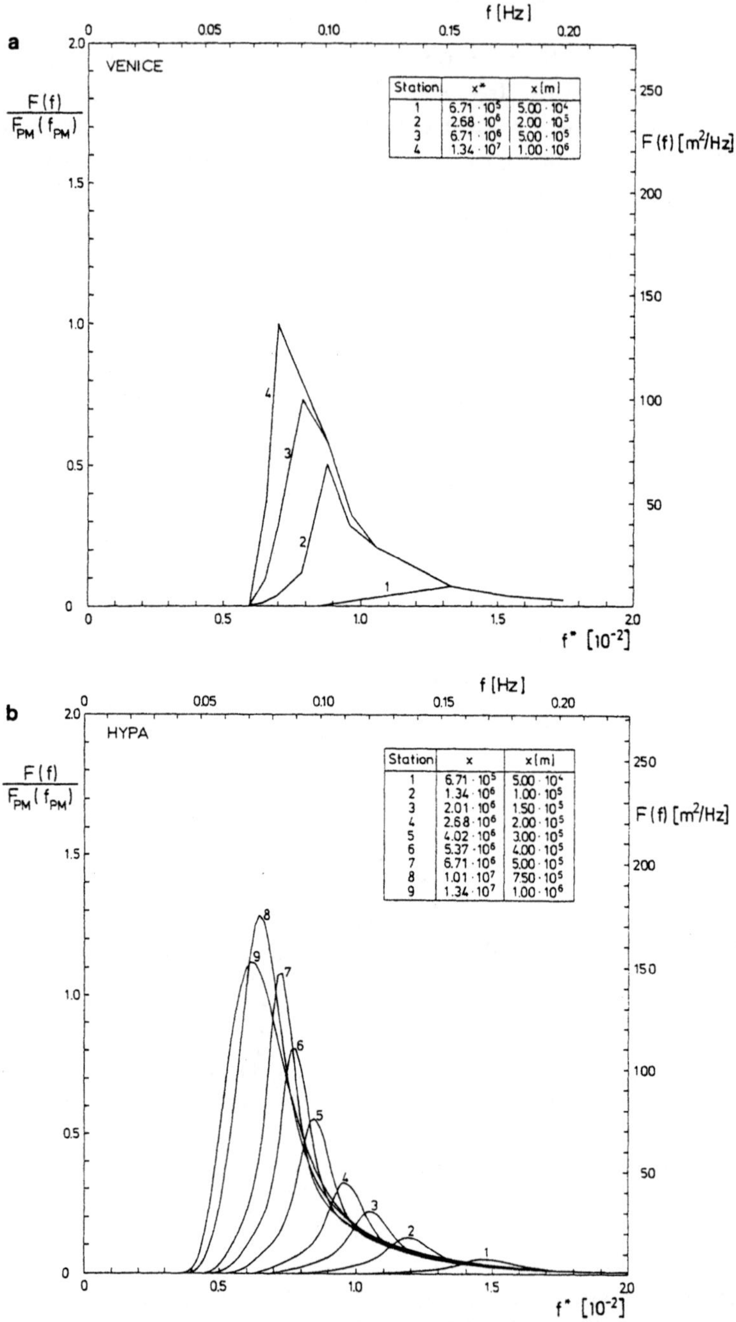

Fig. 7.3. Examples of fetch-limited spectral growth curves, Case II, for model classes DP (VENICE, panel a), CH (HYPA, panel b), CD (SAIL, panel c), and EXACT-NL (panel d). Note the lack of overshoot in the DP spectra and the noisy appearance of the EXACT-NL spectra.

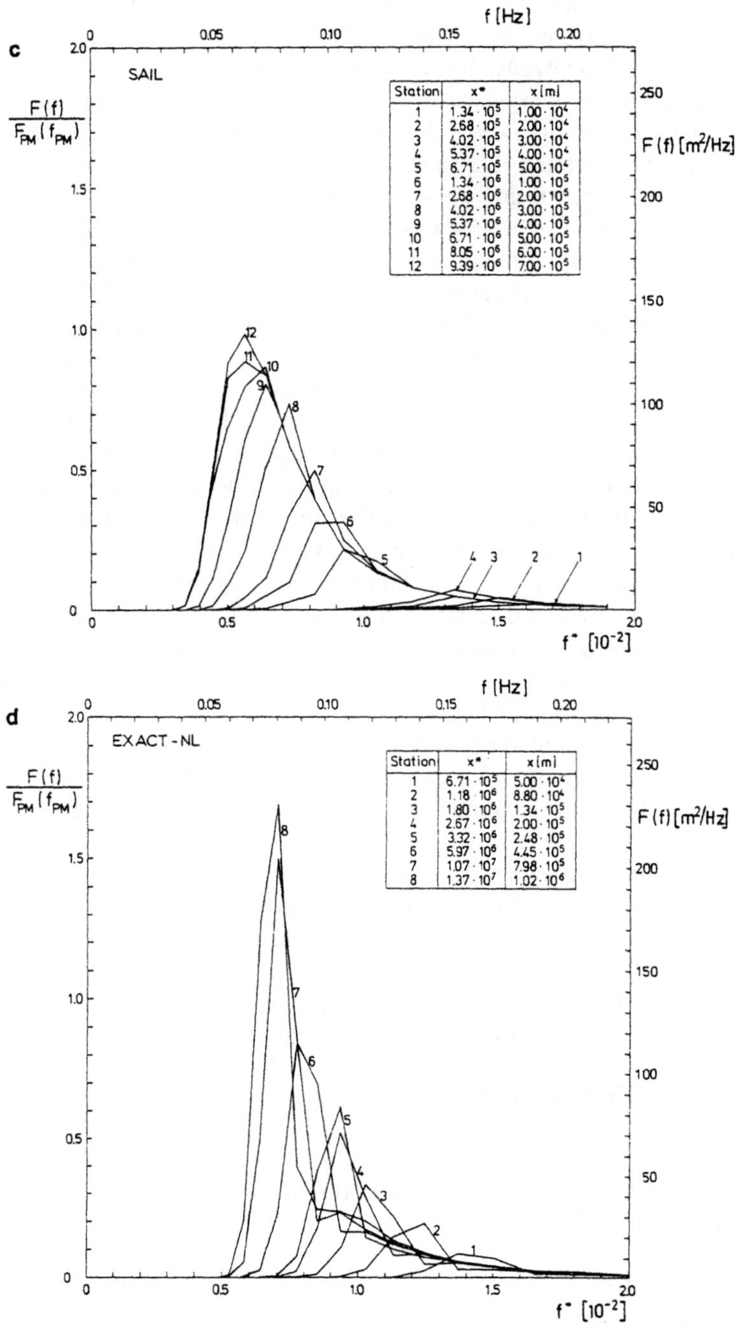

Fig. 7.3. *(continued)*

which wave measurements are most frequently performed. However, the data of Mitsuyasu *et al.* (1980) suggest that wave spectra can indeed sometimes develop beyond the Pierson–Moskowitz form.

The growth of the spectra as a function of fetch is shown in Fig. 7.3 for three typical representatives of the three model classes and for the EXACT-NL model. First-generation DP models (exemplified by VENICE, Fig. 7.3a) exhibit nested spectral growth and are unable to reproduce the overshoot phenomenon, which is clearly seen in the second-generation models HYPA, class CH (Fig. 7.3b), and SAIL, class CD (Fig. 7.3c), and in EXACT-NL, class CD (Fig. 7.3d). The spectral growth curves for the CH and CD models are very similar (in fact, all three have been calibrated with the same fetch-limited JONSWAP growth data). The last development stage of the HYPA model, showing a decrease in the peak spectral energy density, is a peculiar feature of this model and does not appear in the other hybrid models. Available data on the transition to a fully developed stage are inadequate to determine whether this feature is realistic. It appears in HYPA (without being specifically intended) through the particular form chosen for the prognostic equations for the spectral shape parameters, which are kept fixed in the other CH models. The noisy structure of the EXACT-NL spectra cannot be attributed to numerical integration instabilities, but appears to represent an inherent feature of the exact nonlinear energy transfer process. Observed fetch-limited spectra show a similar variability in excess of that expected from statistical sampling fluctuations.

The duration-limited one-dimensional spectral growth curves are generally similar to the fetch-limited growth curves, except that the corresponding energy and frequency parameters change at a different rate, as discussed subsequently.

For all models except SAIL and EXACT-NL, the associated directional distributions in the saturation region of the spectrum are taken as constant factors which are independent of frequency or the stage of windsea development (normally proportional to $\cos^2\theta$ or $\cos^4\theta$ in the downwind half-plane and zero in the upwind half-plane, cf. Chapters 4–6). SAIL assumes a Mitsuyasu *et al.* (1975) directional spreading function dependent on f/f_{PM}, while the directional distribution of EXACT-NL is not prescribed, but is computed explicitly from the energy balance equation.

7.4. FETCH-LIMITED AND DURATION-LIMITED GROWTH CURVES

Figures 7.4 and 7.5 show the nondimensional energy E^* and peak frequency f_p^*, respectively, for all models as a function of fetch x^* for large t^* (stationary state). The corresponding duration-limited growth curves for

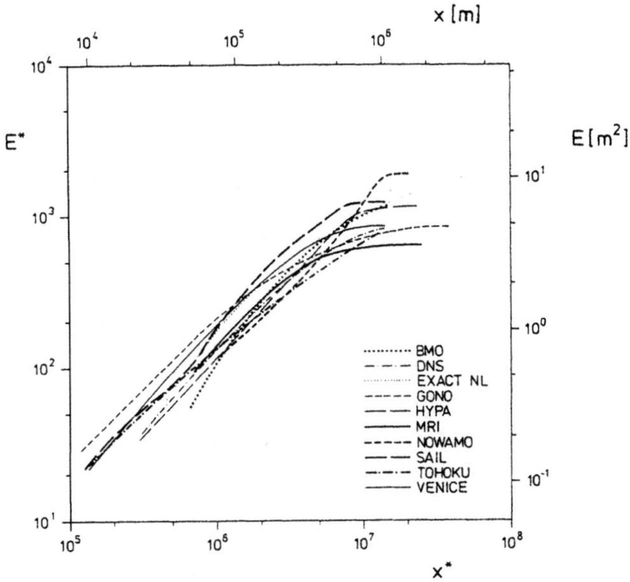

Fig. 7.4. Nondimensional fetch-limited growth curves for the total energy E^* (Case II).

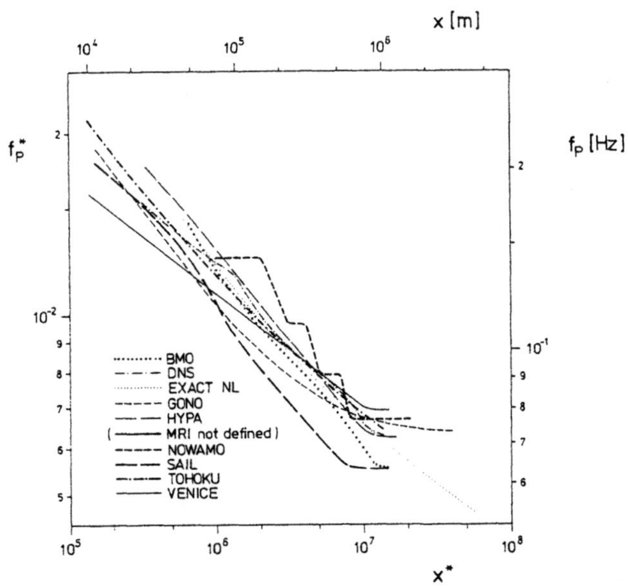

Fig. 7.5. Nondimensional fetch-limited growth curves for the peak frequency f_p^* (Case II).

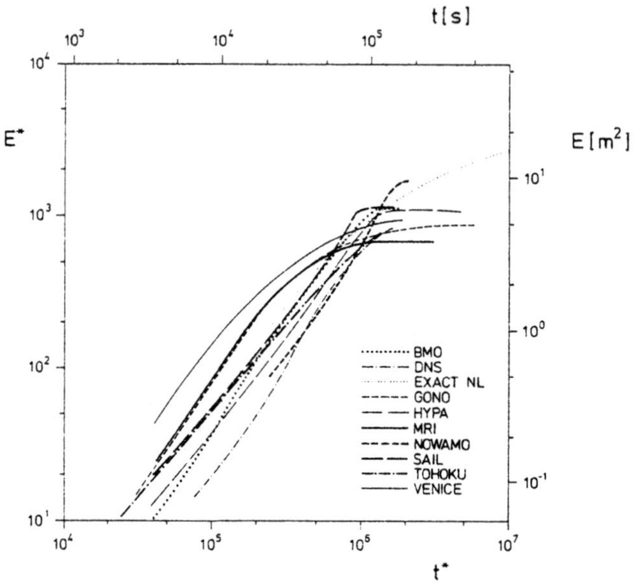

Fig. 7.6. Nondimensional duration-limited growth curves for the total energy E^* (Case II).

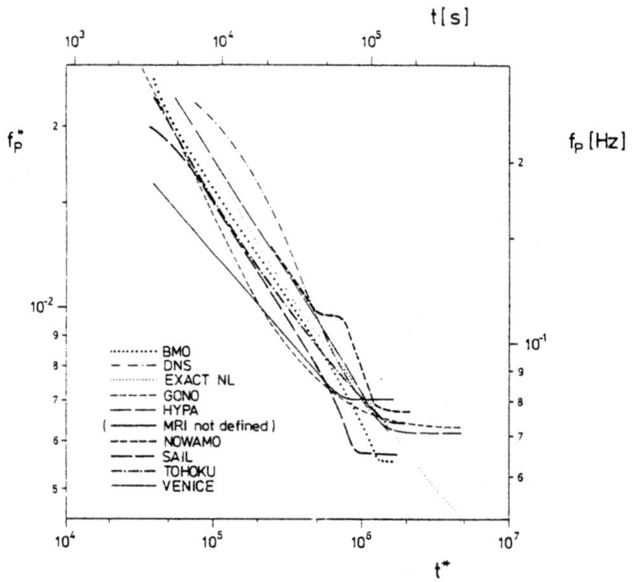

Fig. 7.7. Nondimensional duration-limited growth curves for the peak frequency f_p^* (Case II).

infinite fetch (in practice, $x = 1000$ km in most cases) are shown in Figs. 7.6 and 7.7.

The strong differences apparent between different models in Figs. 7.4–7.7 are striking, since all models were calibrated against fetch- or duration-limited data.

As pointed out in Chapter 3, much of this scatter can probably be attributed to the indeterminacy of the drag law relating u_* to U_{10}, and the uncertainty as to whether u_* or U_{10} is the more appropriate variable for describing wave growth. It is also possible that additional parameters, such as the gustiness of the wind or the background turbulence in the ocean, contribute to the observed differences in wave growth rates (cf. Kahma, 1981; Komen *et al.*, 1985). To these difficulties must be added the problem of relating near-surface wind or friction velocities via a boundary layer model to the geostrophic winds provided by normal atmospheric models as input for the wave models.

We have not attempted to resolve the origins of the differences in the resultant model wave growth rates here. However, we may attempt to separate the discrepancies associated with the atmospheric input from the more fundamental differences related to the internal wave dynamics of the models. Since the wind field is uniform and stationary throughout the experiment, any differences in the formulation of the wave input in terms of the prescribed wind speed at 10 m height can be expressed by a constant factor modifying the effective drag coefficient (after the growth rate has been expressed in terms of u_* for all models). We may therefore eliminate all discrepancies which can be attributed to differences in the formulation of the effective driving wind stress by redefining the drag coefficient for each model to yield fetch-limited growth curves which lie as close as possible to some given mean curve. We have chosen the central portion of the observed JONSWAP growth curves for this purpose. [These agree closely, for suitably chosen drag coefficients, with the growth curves of Mitsuyasu *et al.* (1971) and other experimenters. It should be noted, however, that the JONSWAP growth curves have been translated here from the measured $\tilde{x} = gx/U_{10}^2$ values used in Hasselmann *et al.* (1973) into an equivalent x^* dependence using the present drag coefficient $c_d = 1.83 \times 10^{-3}$. The logarithmic x^* axes also shown in the JONSWAP growth figures were based on a drag coefficient $c_d = 1.0 \times 10^{-3}$ and should therefore be displaced relative to the logarithmic $\tilde{x}$ axes to agree with the present definitions.]

Table 7.1 lists the ratios c_d'/c_d of the modified drag coefficient c_d' relative to the original drag coefficient $c_d = 1.83 \times 10^{-3}$. The scatter of the ratios c_d'/c_d is not inconsistent with the range of variability found in field measurements of c_d for a given wind speed.

The renormalized growth curves are shown in Figs. 7.8–7.11. The directions in which individual curves are displaced by the renormalization

Table 7.1
Ratios of Drag Coefficient c_d' Used in Renormalized Figs. 7.8–7.11 to Original Drag Coefficient c_d Used in Figs. 7.4–7.7

Model	c_d'/c_d (from $E^* - x^*$ growth curves)
MRI	1.05
VENICE	1.44
NOWAMO	0.96
GONO	1.30
TOHOKU	1.00
HYPA	1.00
BMO	1.04
SAIL	1.52
DNS	1.00
EXACT-NL	1.17

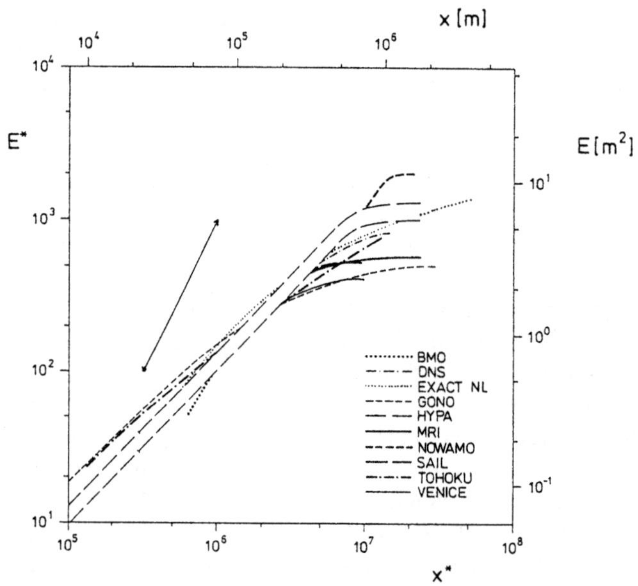

Fig. 7.8. Renormalized fetch-limited growth curves for the total energy E^*. The white band represents the JONSWAP growth curve (same as HYPA) $\pm 5\%$. The arrows indicate the direction along which the curves are shifted by redefining c_d (Case II).

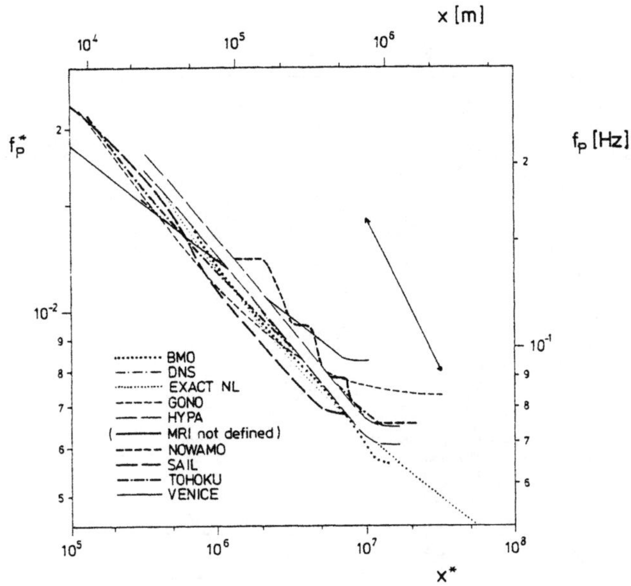

Fig. 7.9. Renormalized fetch-limited growth curves for the peak frequency f_p^*. (See caption for Fig. 7.8.) (Case II).

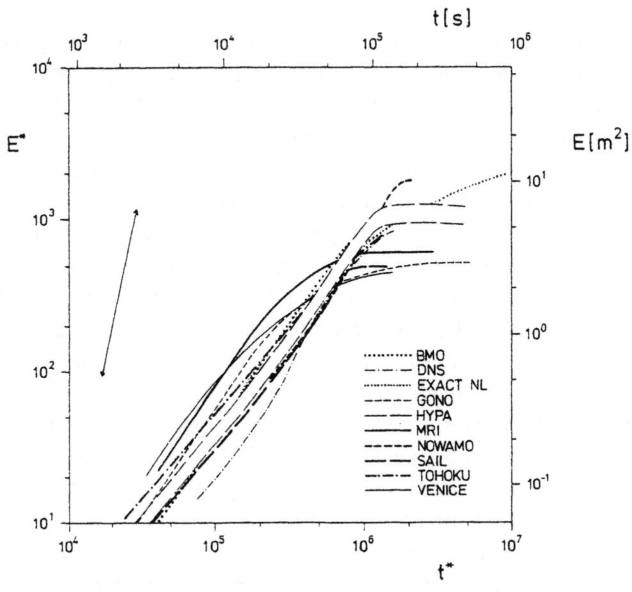

Fig. 7.10. Renormalized duration-limited growth curves for the total energy E^*. (See caption for Fig. 7.8.) (Case II).

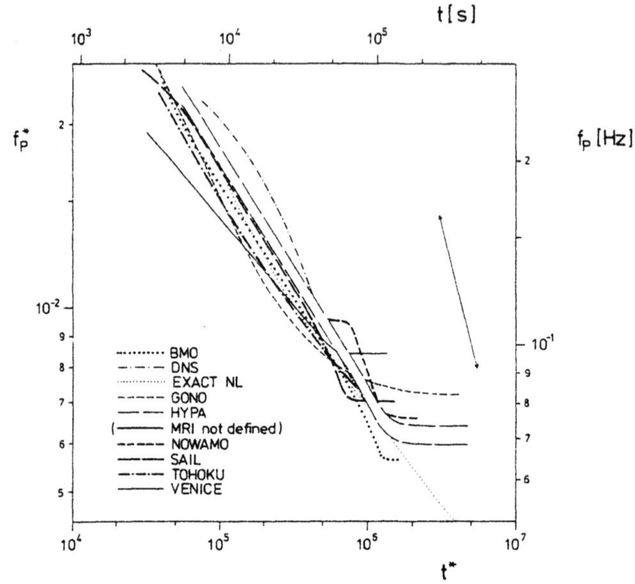

Fig. 7.11. Renormalized duration-limited growth curves for the peak frequency f_p^*. (See caption for Fig. 7.8.) (Case II).

are indicated by arrows. It is seen that while the fetch-limited growth curves for E^*, Fig. 7.8, now show much greater agreement, significant discrepancies still remain in the other growth curves. It is clearly not possible, with a single redefinition of c_d, to achieve good agreement simultaneously in the growth curves for both fetch- and duration-limited growth for both scale parameters E^* and f_p^*. It is also not always possible to achieve good agreement simultaneously with respect to the slopes and the asymptotic levels of the growth curves. Basic discrepancies in the growth laws for the spectral distributions, for example the lack of an overshoot in the DP spectral growth curves, also remain unaffected by the renormalization.

However, the renormalized growth curves now exhibit a more distinct clustering into model classes. In particular, duration-limited growth for the first-generation DP wave models is seen to be too large by a factor of order 1.5 relative to the second-generation models. As pointed out in Chapters 4 and 5, this is a fundamental shortcoming of DP models which cannot be remedied by modifying the source functions.

The renormalized second-generation wave models show reasonable mutual agreement. However, this is not surprising. The EXACT-NL model was tuned, for example, after prescribing the input and exact nonlinear transfer source functions S_{in} and S_{nl}, to reproduce the observed fetch-limited growth curves through appropriate choice of the dissipation source function S_{ds}. The parameterization of the nonlinear transfer in the CD models and the growth rates of the CH models were similarly tuned to fetch-limited data. After renormalization to a common fetch-limited growth rate, the agreement for duration-limited growth follows automatically to first order, provided the parameterization of the nonlinear transfer is scaled correctly in accordance with the scaling of the exact expression. A more sensitive test of the parameterization of the nonlinear transfer in second-generation models is provided by the turning wind Case VII, in which the spectral distribution deviates from the directionally symmetric windsea distributions used to develop the parameterizations.

7.5. CONCLUSIONS AND RECOMMENDATIONS

1. Many of the differences found in the model growth curves may be attributed to the uncertainty as to whether the friction velocity u_*, the wind velocity U_{10}, or possibly other boundary layer parameters control the rate of growth of waves, compounded by the uncertainty in the drag coefficient c_d relating the friction velocity u_* to 10 m wind speed U_{10}.
2. A similar source of uncertainty is the relation between the near-surface velocities u_* and U_{10} and the geostrophic velocities normally provided as input by atmospheric models driving the wave models.
3. These discrepancies could be identified and reduced by reanalyzing already existing wave growth data in conjunction with drag coefficient data obtained under well defined sea state conditions. A joint analysis of such data would provide more consistent empirical wave growth and drag coefficient relations by simultaneously optimizing the fit of both relations to a combined data set. The drag coefficient relation would need to be formulated in this case as a dimensionless expression dependent not only on wind speed and atmospheric stability, but also on sea state and the stage of development of the wind sea (cf. Donelan, 1982; Kitaigorodskii and Zaslavsky, 1974; Liu and Ross, 1980).

   Agreement on a common "standard growth law" obtained in this manner (subject to subsequent revisions) would greatly facilitate future wave model intercomparison exercises. An analysis of the residual scatter of individual drag coefficient and wave growth data about "standard" mean relations would also provide useful background noise data for interpreting the significance of model deviations.

4. Appreciable discrepancies between the growth curves of first- and second-generation models remain even after renormalization of the growth rates to eliminate discrepancies associated with the uncertainties of the wind input. These represent genuine differences in the dynamics of quasi-linear decoupled and nonlinear coupled propagation models which cannot be removed by modifying individual source functions.

Chapter 8

Case III: Slanting Fetch

8.1. WIND FIELD, BOUNDARY CONDITIONS

The geometry of the slanting fetch test case is shown in Fig. 8.1. An initially ($t=0$) calm ocean of dimension approximately 1000 km $\times$ 1000 km was subjected to a 20-m/s uniform wind directed diagonally (45°) across it. The x and y axis directions are defined as 90° and 0°, respectively (angles are measured clockwise from "north"). The wave field was zero on the upwind boundaries $x=0$ and $y=0$; the downwind boundaries at $x=1000$ km and $y=1000$ km were open. Model simulations of the space/time behavior of the resulting wave field were recorded at selected locations noted as A–F in Fig. 8.1. In addition, certain integral properties of the wave field were presented over the entire domain. Here we concentrate only on the steady-state behavior of the models, as it illustrates the most interesting results of the test.

8.2. PURPOSE OF THE TEST

The main emphasis of the test was on the directional response of the wave field to an asymmetrical boundary condition. Thus the discussion will focus on the two-dimensional spectra at a relatively asymmetric location in the test region, point F: $(x,y)=(750 \text{ km}, 75 \text{ km})$, $(x^*, y^*)=(1.0\times10^7, 0.1\times10^7)$. The general spatial distribution of the wave field will be discussed in terms of selected integral properties (total energy, peak frequency, and mean direction).

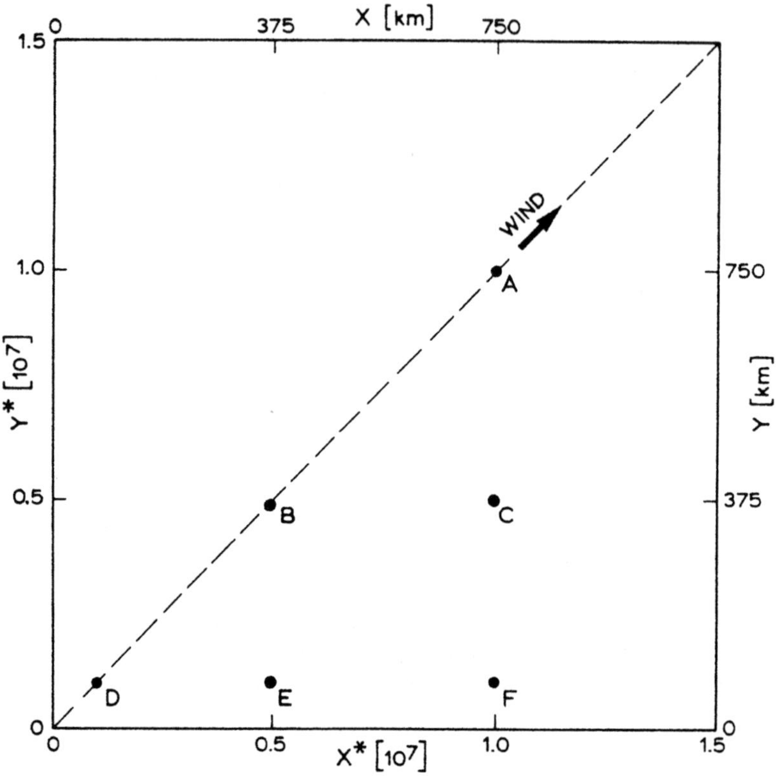

Fig. 8.1. Wind field geometry for Case III, slanting fetch. Special output points are denoted A–F. Two-dimensional spectra are shown in Fig. 8.4 for the point F, $(x, y) \approx (750$ km, 75 km$)$, $(x^*, y^*) \approx (1.0 \times 10^7, 0.1 \times 10^7)$.

Although the wind field is relatively simple, the 2d wave spectrum is governed by a balance involving the directionally dependent wind input; advection, particularly in directions nonparallel to the wind; nonlinear coupling; and dissipation. The case thus tests the interplay of a number of processes under asymmetrical generation conditions for which the models were not originally tuned.

8.3. RESULTS

The spatial distributions of the nondimensional energy E/E_{PM}, nondimensional peak frequency f_p/f_{PM}, and mean wave direction $\bar{\theta}$ are shown in Figs. 8.2 and 8.3 for two selected models in the form of isoline plots of

f_p/f_{PM} and Custer diagrams, consisting of fields of arrows of length E/E_{PM} and direction $\bar{\theta}$ (named in memory of General Custer's legendary encounter with overwhelming Indian forces at the battle of Little Big Horn).

Table 8.1 provides a summary of the nondimensional energies and frequencies of all models at three selected locations A, B, and C (cf. Fig. 8.1), while Table 8.2 lists the same data, but with the energies and frequencies normalized individually by the values predicted by the model for Case II, for the same parallel fetch. The parallel fetch is defined here as the distance upwind to the coast, measured parallel to the wind direction. Normalization with respect to Case II emphasizes the relative effects on each model of the wind field geometry, while eliminating the effects of different fetch-limited growth rates for different models which were discussed in Case II.

The set of nine panels in Fig. 8.4 shows the 2d spectral distributions for all models (except EXACT-NL) at the point F close to the coast, for which the effects of asymmetrical wave growth may be expected to be particularly pronounced (cf. Fig. 8.1). The differences between the models at this point are summarized in Fig. 8.5, which shows the relative positions of the models in the parameter phase plane E_{III}/E_{II} versus f_{pIII}/f_{pII}. The mean wave direction $\bar{\theta}$ is indicated in the figure by a (unit length) arrow.

The following general conclusions may be drawn from the figures and tables:

1. Strong effects of the asymmetrical growth geometry away from the central diagonal are seen in the two DP models MRI and VENICE. The asymmetry arises because wave components which propagate in different directions on either side of the mean wind direction are exposed to different fetches. Since in DP models each wave component evolves independently, this leads to strongly skewed directional distributions off the diagonal (Figs. 8.3a, 8.4a, b).

2. These models also experience a reduced growth rate relative to Case II along the diagonal, because waves traveling at an angle to the wind experience a smaller fetch than in the orthogonal fetch geometry (cf. Table 8.2). The effect is more pronounced for the model MRI than for the VENICE model, as the MRI directional distribution is broader than that of the VENICE model.

3. A skewed directional distribution at points away from the diagonal is also found in two of the three CD models, SAIL and DNS (Figs. 8.4h, i, 8.5). This is because these models also keep track of the growth history of individual windsea components, while at the same time allowing for nonlinear coupling between the components.

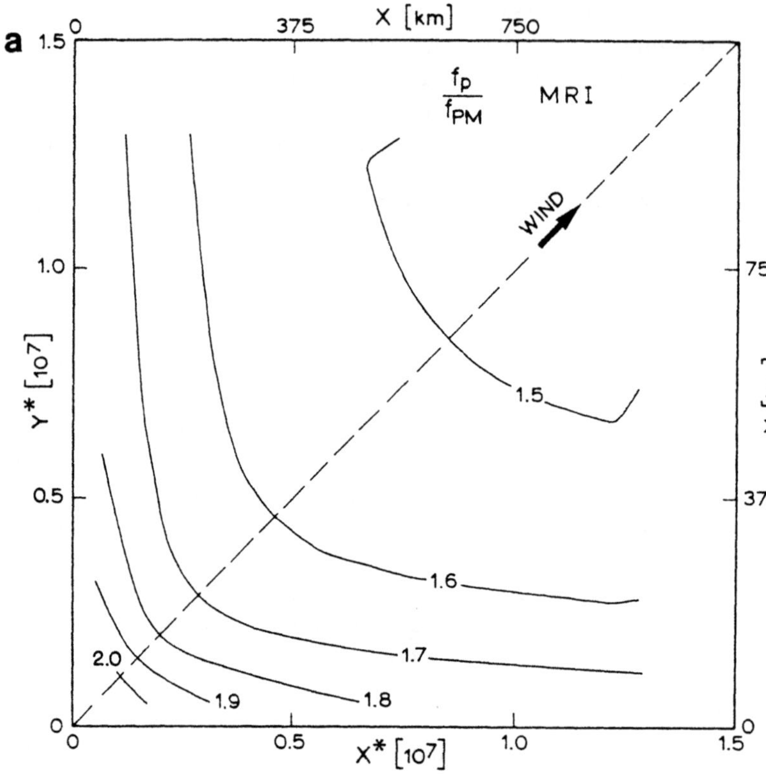

Fig. 8.2. Isolines of f_p/f_{PM} for the stationary slanting fetch Case III for the DP model MRI (panel a) and the CH model TOHOKU (panel b). The CH model exhibits less influence of the wind field asymmetry (isolines parallel to the coast).

4. For the same reason these models also experience a reduced growth rate along the diagonal (Table 8.2). This effect is particularly marked for SAIL at short fetches (point B), since this model employs a Mitsuyasu-type directional distribution dependent on the stage of development of the windsea, which is particularly broad at small fetches.

5. The one-parameter CH models NOWAMO and TOHOKU show no directional skewing away from the central diagonal, as the local windsea direction is restrained to coincide with the local wind direction (cf. Figs. 8.3b, 8.4c, e). As the growth rates for these models depend only on the parallel fetch (cf. Fig. 8.2b), they should show

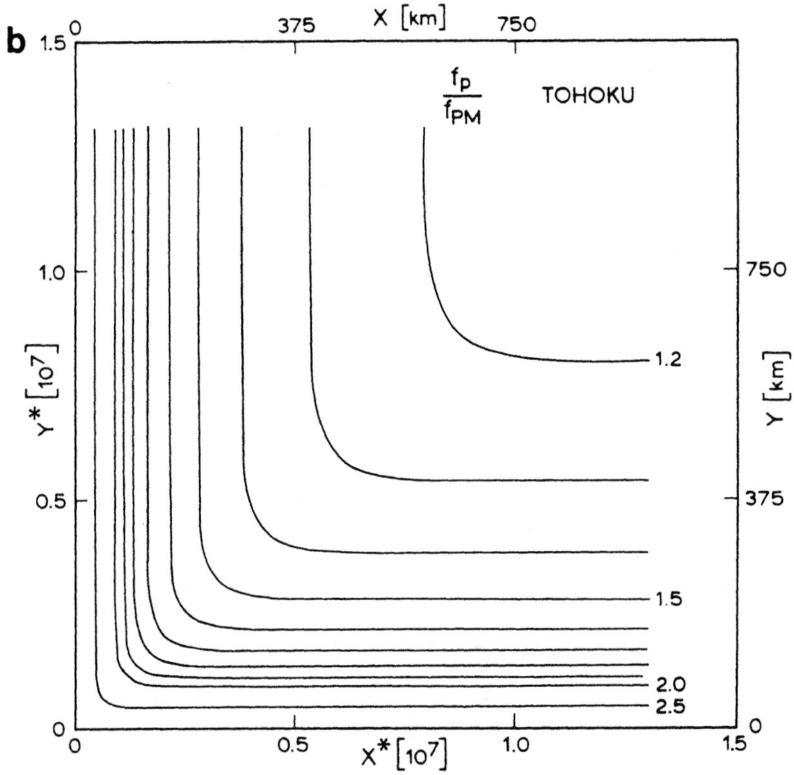

Fig. 8.2. (*continued*)

no decrease in the growth rate relative to Case II. The small reductions seen in Table 8.2 can probably be attributed to some unavoidable numerical diffusion in the advection schemes which leads to some dependence on the cross-wind variations of the wave energy.

6. The third CD model, BMO, similarly constrains the direction of the windsea spectrum to the local wind direction, and therefore also exhibits no directional skewness (cf. Fig. 8.4g). However, in contrast to most one-parameter CH models, the adaptation of the windsea spectrum to a prescribed form is carried out in this CD model after the incremental changes of the individual spectral components due to advection have been evaluated. Thus cross-wind gradients of the spectrum can indirectly affect the mean downwind growth rate, and

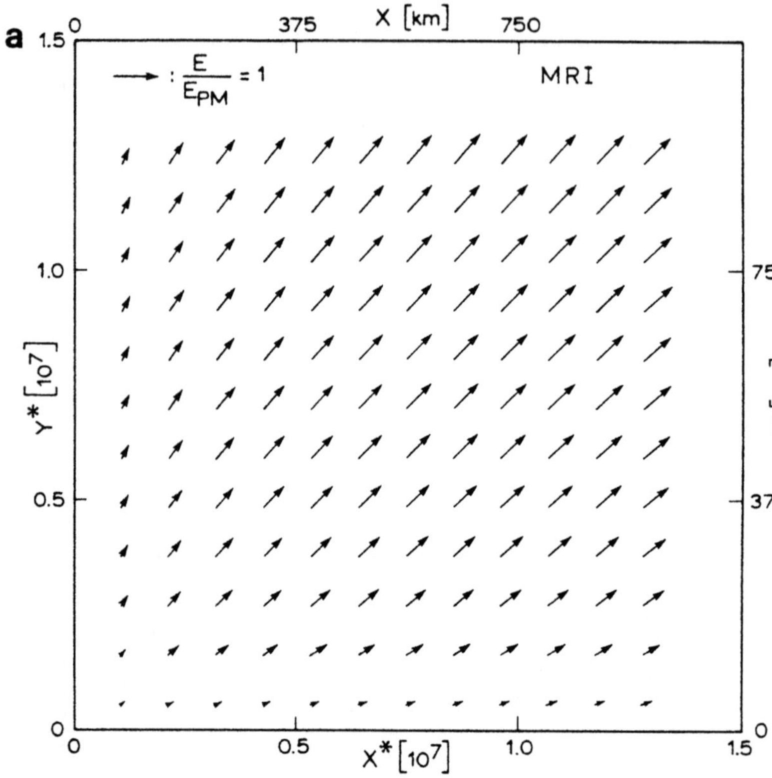

Fig. 8.3. Custer diagrams for the slanting fetch Case III for the DP model MRI (panel a) and the one-parameter CH model TOHOKU (panel b). (Note scales for the vector length.) The mean wave direction is skewed toward the coast for the DP model, but not for the one-parameter CH model. Custer diagrams similar to panel a are found for all DP and CD models (except BMO) and the multiparameter model HYPA, while panel b is typical for BMO and the one-parameter models NOWAMO and GONO (although the latter shows a weak 10° skew away from the wind direction near the coast).

the resultant downwind growth rates are therefore smaller than in Case II (cf. Table 8.2).

7. The remaining one-parameter CH model, GONO, bears some similarity to the BMO model in the sense that it also constrains the mean windsea direction to the local wind direction, but considers the incremental change in the mean windsea direction induced by the advection of wave energy. If this differs from the change required by

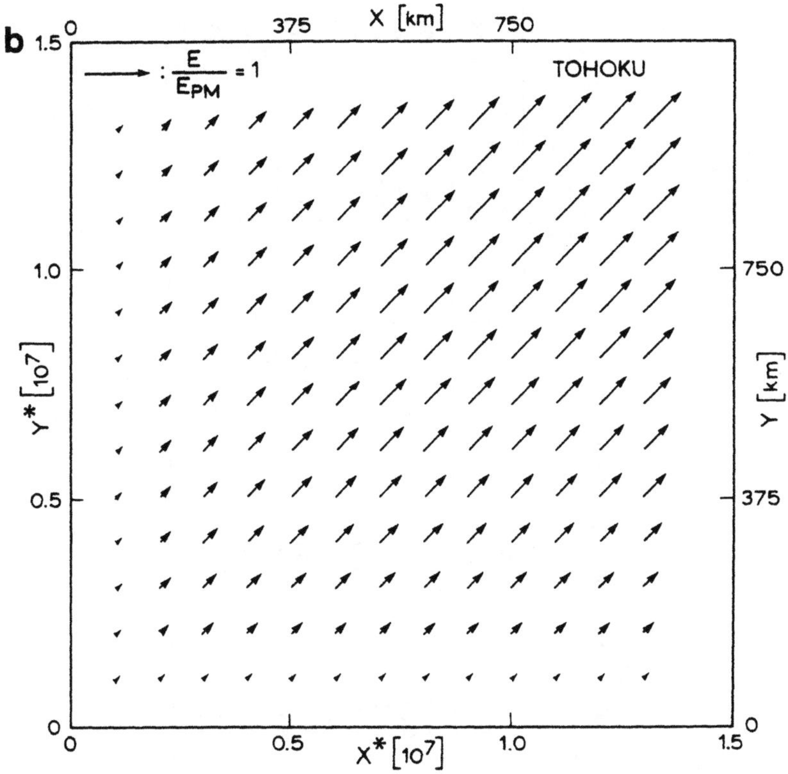

Fig. 8.3. (*continued*)

the wind directional constraint, the adjustment to the wind direction is assumed to be accompanied by some energy loss (cf. Table 8.2). The small asymmetry apparent in the GONO spectrum at point F, Fig. 8.4d, arises because the model picks up some low-frequency swell at angles beyond 90° in addition to the dominant local windsea at 45°.

8. The multiparameter CH model HYPA, finally, attempts a compromise between the discrete models, which have sufficient degrees of freedom to simulate detailed, frequency-dependent directional windsea distributions, and the one-parameter models, in which the windsea directional distribution is determined by the local wind direction. This model is able only to predict the mean directional skew of the wave field away from the diagonal (its Custer diagram is

Table 8.1
Energy and Frequency Ratios for Selected Points—Case III/PM

		E_{III}/E_{PM}			f_{pIII}/f_{PM}		
Model $(\Delta x, \Delta t)$[a]	Position: (x, y) (km)	B (375, 375)	C (750, 375)	A (750, 750)	B (375, 375)	C (750, 375)	A (750, 750)
MRI (40,60)		0.45	0.50	0.59	1.59	1.54	1.48
VENICE (75,30)		0.53	0.60	0.71	1.39	1.34	1.25
NOWAMO (75,90)		0.62	0.71	1.34	1.94	1.70	<1.6
GONO (75,90)		0.57	0.60	0.78	1.30	1.30	1.19
TOHOKU (25,30)		0.45	0.48	0.70	1.37	1.32	1.05
HYPA (25,15)		0.59	0.71	0.90	1.33	1.30	1.07
BMO (25,20)		0.63	0.70	0.90	1.25	1.20	0.91
SAIL (37.5,60)		0.53	0.85	1.00	1.25	1.13	0.95
DNS (50,20)		0.45	0.48	0.72	1.40	1.34	1.25
JONSWAP (N/A)		0.57	0.90	1.23	1.14	1.02	0.91

[a] $(\Delta x, \Delta t)$ refer to model grid spacing (km) and time step (min).

Table 8.2
Energy and Frequency Ratios for Selected Fetches—Case III/Case II

Model		E_{III}/E_{II}			f_{pIII}/f_{pII}		
	Position:	B	C	A	B	C	A
	Fetch:	(375, 375)	(750, 375)	(750, 750)	(375, 375)	(750, 375)	(750, 750)
MRI		0.88	0.99	0.97	1.03	1.00	1.00
VENICE		0.91	0.86	0.95	0.98	1.01	0.98
NOWAMO		—	1.33	1.30	0.76	0.80	0.84
GONO		0.90	0.83	1.04	1.05	1.09	1.02
TOHOKU		0.92	0.75	0.95	1.04	1.10	0.91
HYPA		0.84	0.73	0.91	1.04	1.12	0.96
BMO		0.83	0.73	0.90	0.97	1.17	0.91
SAIL		0.56	0.85	1.00	1.23	0.96	0.84
DNS		0.82	0.70	0.96	1.09	1.13	1.11

similar to that of MRI, Fig. 8.3a). The 2d spectrum at point F (Fig. 8.4f) accordingly exhibits a mean wave direction turned coastward by about 20° relative to the wind direction. However, since the model still contains only one prognostic directional parameter, it yields a 2d spectrum which is constrained to a symmetrical, frequency-independent distribution about the mean wave direction, and is unable to simulate the frequency-dependent skewed directional distributions typical of models which apply no constraints to the growth region of the windsea spectrum (Figs. 8.4a, b, h, i).

9. Despite the differences in the detailed 2d distributions, the integrated spectral parameters for point F in Fig. 8.5 show relatively little divergence (after normalization with respect to Case II). As expected, models which turn the wave direction parallel to the coast tend to generate lower-frequency wave components. The weak effects in Fig. 8.5 are consistent with the relatively small changes found for most points in Table 8.2.

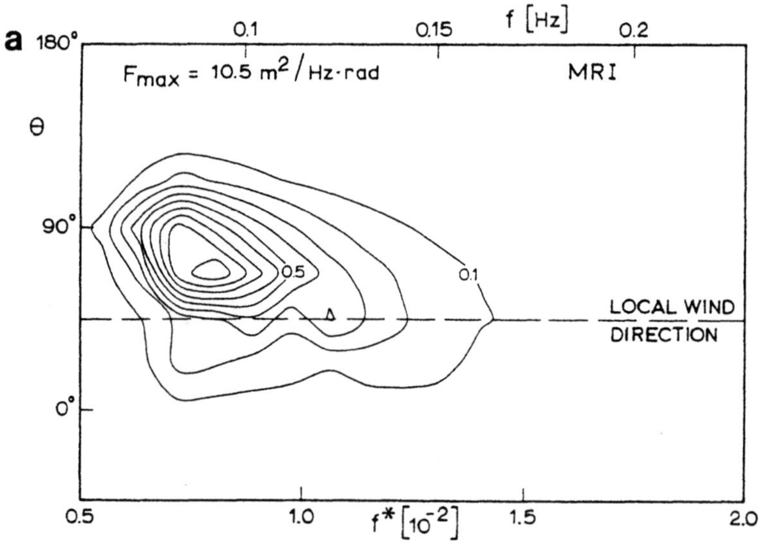

Fig. 8.4. Isoline plots of the 2d spectral distributions at point F (Case III, Fig. 8.1) for the nine operational models of this study. Contours are multiples of 0.1 in units of F/F_{max}. Note the directional fine structure of the DP and CD models (except BMO) relative to the largely prescribed directional distributions of the CH models. The asymmetry seen in the GONO spectrum (panel d) is due to low-frequency swell with directions greater than 90°. The mean windsea direction for HYPA (panel f) is skewed toward the coast in accordance with the mean wave direction of the DP and CD models, but this multiparameter CH model is also unable to simulate the frequency dependence of the windsea directional distribution predicted by the discrete models.

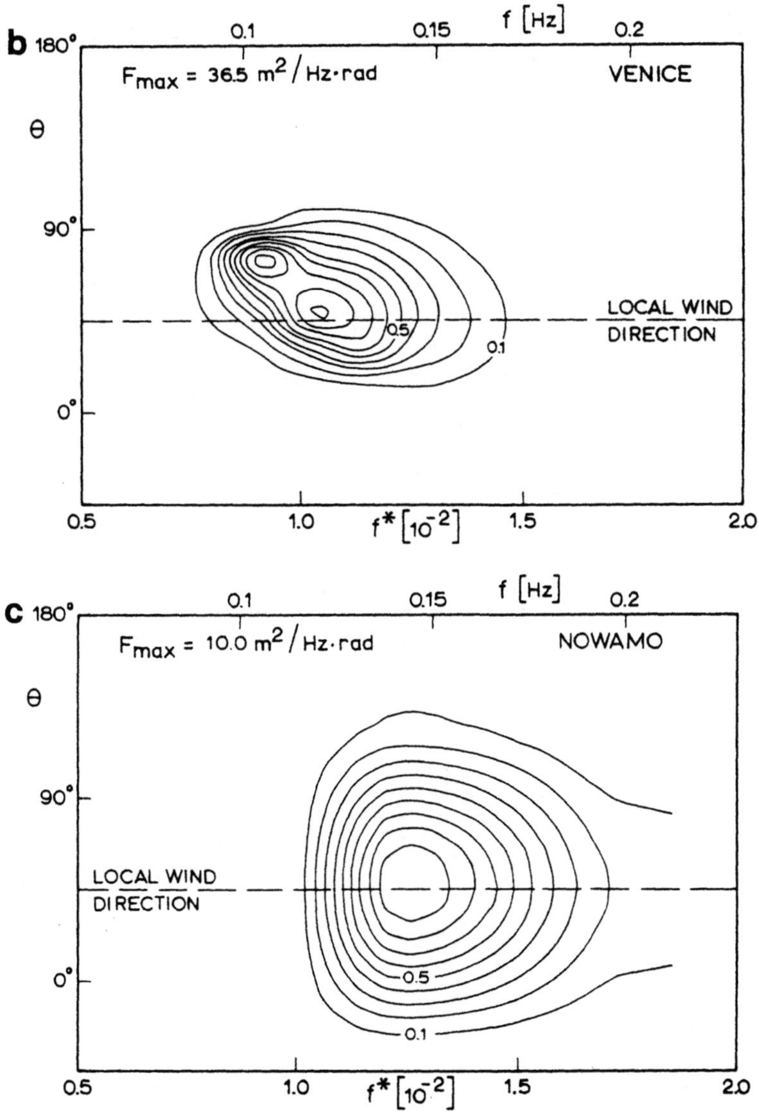

Fig. 8.4. *(continued)*

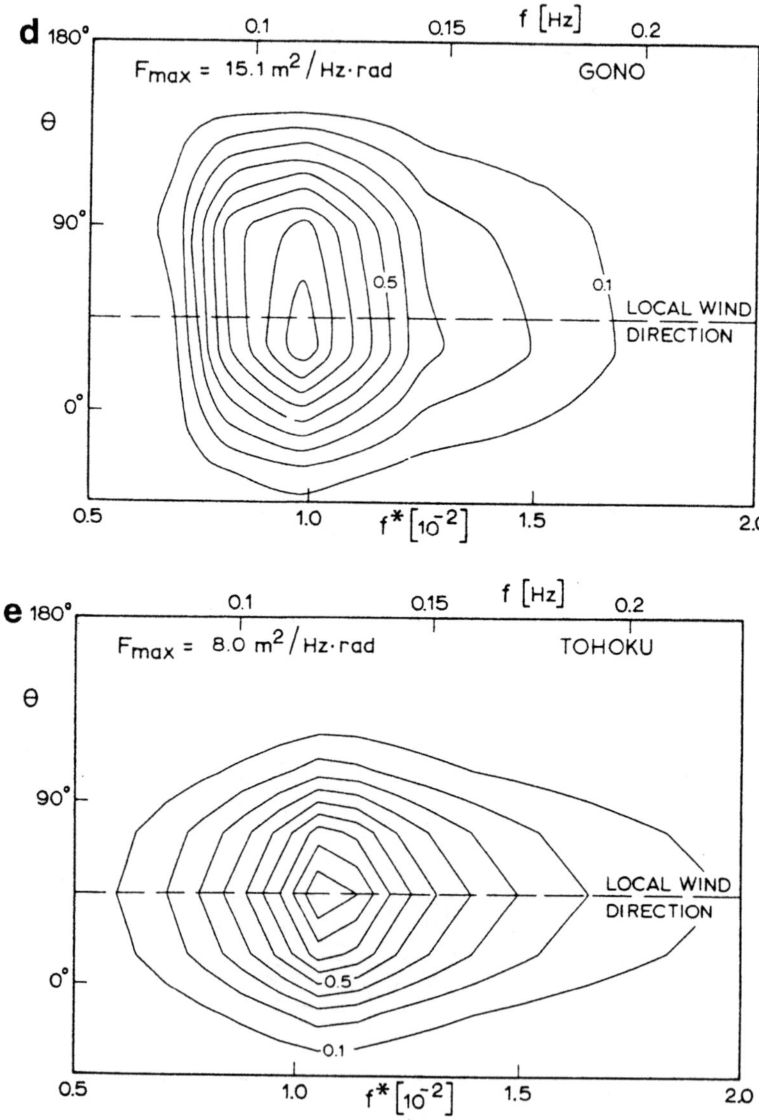

Fig. 8.4. *(continued)*

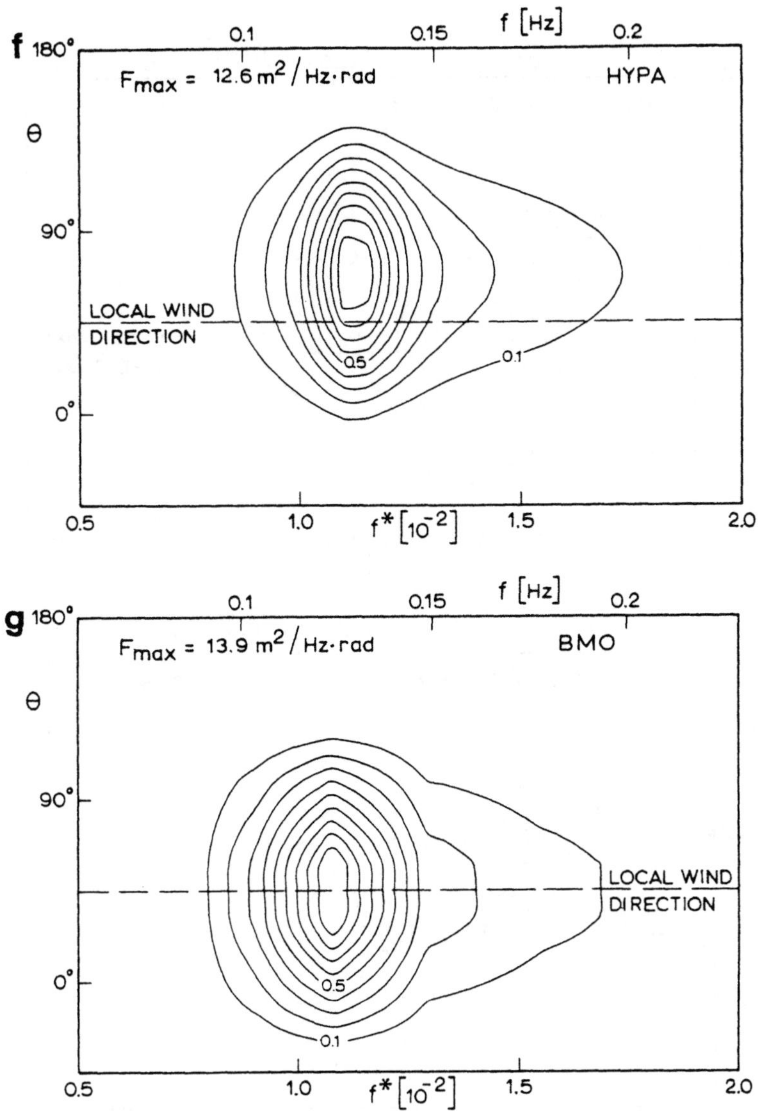

Fig. 8.4. (*continued*)

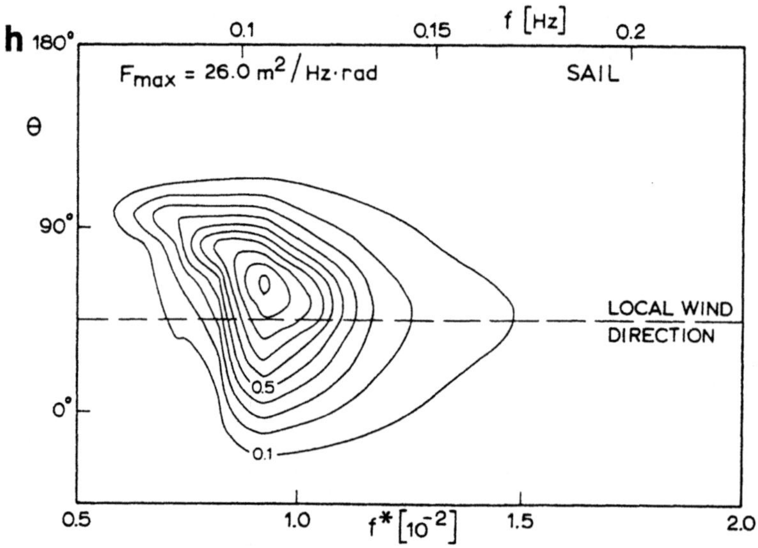

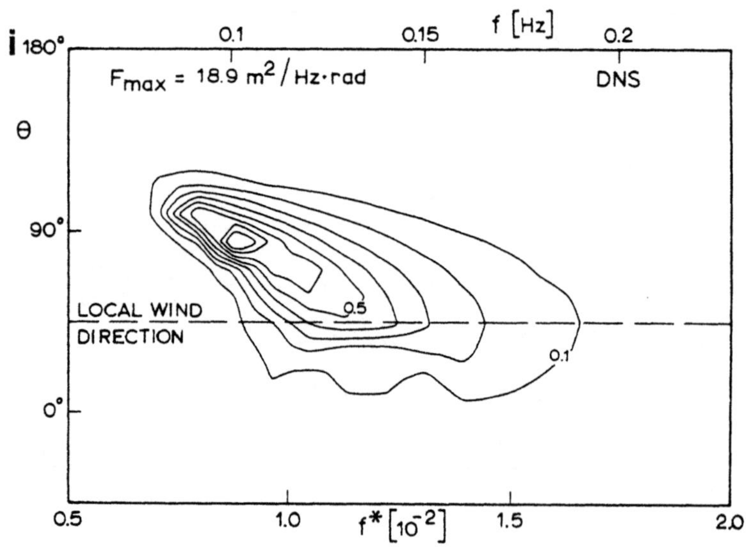

Fig. 8.4. *(continued)*

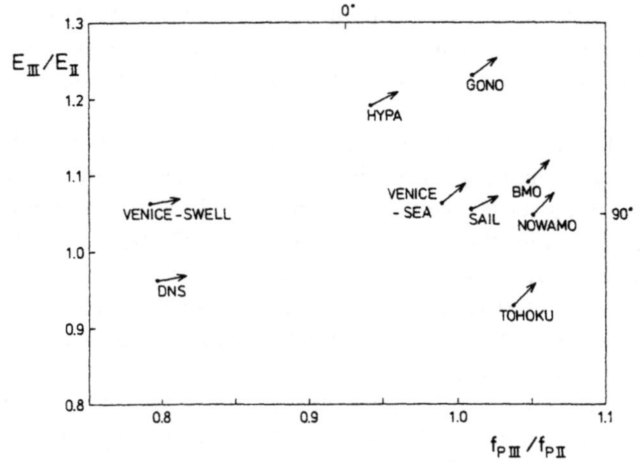

Fig. 8.5. Location of models in the E_{III}/E_{II} versus f_{pIII}/f_{pII} parameter plane at point F (Fig. 8.1, Case III). Unit length arrows indicate mean wave direction (MRI is not shown, as the peak windsea frequency was not well defined for Case II for the small parallel fetch characterizing point F).

8.4. CONCLUSIONS AND RECOMMENDATIONS

For all models, the growth rates of the one-dimensional frequency spectra in the slanting fetch case were slightly retarded but otherwise very similar to the limited-fetch results of Case II (within about $\pm 20\%$) This applied both for growth along the diagonal, where the effective fetch was reduced for waves propagating in directions on either side of the wind direction, and for downwind growth along lines displaced from the symmetry axis, for which the effective fetch was increased for propagation directions on one side of the wind direction and decreased for the other side.

Larger differences were found in the 2d directional distributions. These showed a distinct separation into two classes. The one-parameter CH models TOHOKU, NOWAMO, and GONO and the CD model BMO yielded spatially uniform directional distributions with a mean wave direction parallel to the wind (GONO, however, exhibited some turning of the mean wave direction toward the coast owing to the superposition of an asymmetrical swell field on the symmetrical windsea spectrum). The remaining models yielded 2d windsea spectra weighted toward the directions parallel to the coast, as was to be expected from the fetch asymmetry. The discrete models MRI, VENICE, SAIL, and DNS were able to simulate detailed 2d spectra with frequency-dependent, skewed directional distributions. The multi-parameter hybrid model HYPA was able to reproduce the turning of the mean windsea direction, but since it contained only one prognostic directional parameter, it was unable to capture the details of the frequency dependence and asymmetry of the spreading function about the mean direction.

It should be possible to resolve some of the questions raised by this test case by direct comparison with existing data. Studies of directional distributions for slanting-fetch conditions off the Dutch coast (Holthuijsen, 1983) support qualitatively the skewed directional distributions predicted by the discrete models. It must be recalled, however, that none of the models adequately parameterizes the nonlinear transfer, one of the critical processes governing the skewed directional distributions in the present case, so that the directional fine structures predicted by the discrete models are also questionable (as indicated by the significant differences in detail among the predicted 2d distributions). Further model simulations for this data set and similar data collected elsewhere (e.g., at the JONSWAP site) would be very valuable.

In contrast to some of the more abstract hypothetical wind fields considered in the following test cases, a slanting-fetch wind field clearly represents a commonly occurring wind situation of considerable practical concern, for example for offshore operations. It is therefore noteworthy that this relatively simple case has already exposed several critical elements of the models, particularly with respect to the directional response characteristics.

Chapter 9

Case IV: Half-Plane Wind Field

9.1. WIND FIELD, BOUNDARY CONDITIONS

The geometry of the wind field of this test case is shown in Fig. 9.1. A stationary front, running from south to north, divides the area into two half-planes. A constant wind of $U_{10} = 20$ m/s blows to the north, parallel to the front, over the left half-plane. The right-half plane is calm. There is no energy influx at the leeward boundary and at all other boundaries perfect absorption or the vanishing of energy gradients normal to the boundaries is assumed. The tests were run from an initially calm sea at time $t = 0$ until the model results were stationary.

9.2. PURPOSE OF TEST

The main purpose of this case was to test the windsea–swell transition algorithms of CH and CD models in the radiation of swell from the windy half-plane into the calm half-plane. Also of interest was the influence of the sharp lateral front on the windsea development in the windy half-plane.

9.3. GENERAL CHARACTERISTICS

The main features of the model response in both half-planes can be summarized by Custer diagrams of E/E_{PM} and $\bar{\theta}$. The examples given in Figs. 9.2a–d indicate no marked distinctions with respect to different model types. The DP model VENICE and the CD model DNS exhibit a smoother

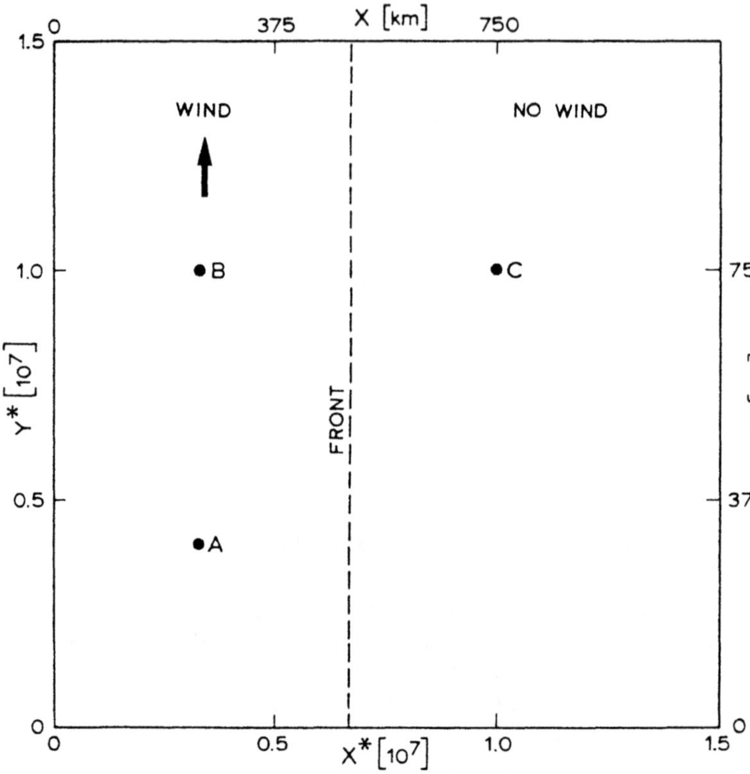

Fig. 9.1. Wind field geometry for Case IV, half-plane wind. A, B, and C denote special output points discussed in the text. In (X^*, Y^*) coordinates, $A = (0.34\times10^7, 0.4\times10^7)$, $B = (0.34\times10^7, 1.0\times10^7)$, $C = (1.01\times10^7, 1.01\times10^7)$.

transition across the front than the CH model HYPA or the hybrid-type model BMO, which, like all hybrid models, suffer a discontinuity in the spectral representation across the front. Otherwise, the energy and direction of the swell field are largely determined by the directional distribution of the windsea in the windy half-plane. Thus the relatively narrow windsea spreading function of VENICE yields a swell field traveling at small angles to the wind and decaying with increasing lateral distance from the front.

In the windy half-plane, the CD model DNS shows the strongest sensitivity to the lateral boundary (in analogy to the sensitivity of this model to the asymmetric boundary discussed in Case III, cf. Figs. 84i and 8.5) while the other models shown in Fig. 9.2 are only weakly affected by the boundary (either because of the narrow directional distribution, in the case

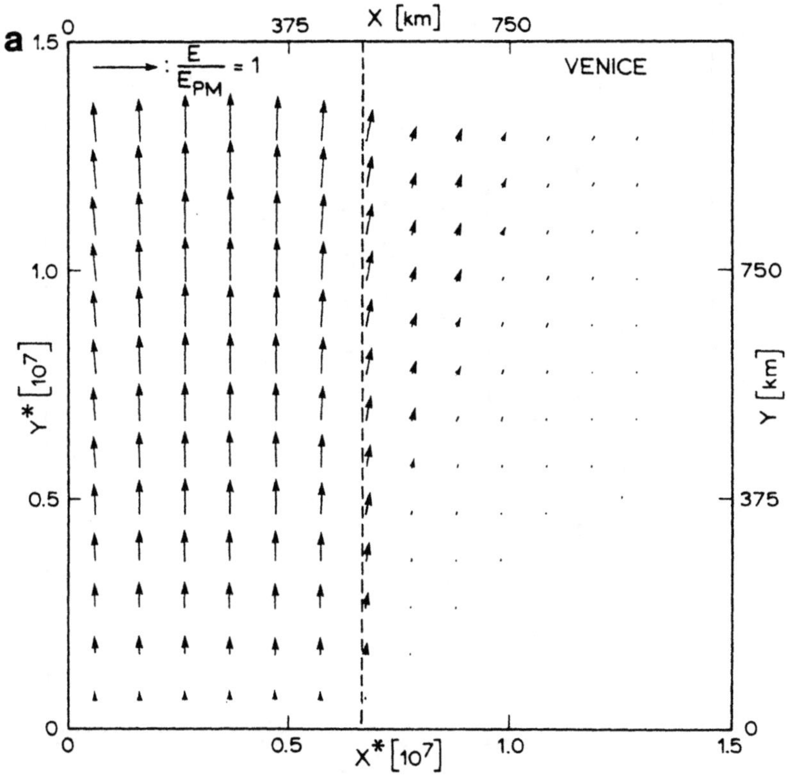

Fig. 9.2. Custer diagrams for Case IV for the DP model VENICE (panel a), the CH model HYPA (panel b), and the CD models DNS (panel c) and BMO (panel d). (Note different vector length scale for DNS.)

of VENICE, or because of the parametrical distribution of the windsea, in the case of the other models shown).

9.4. THE WINDY HALF-PLANE

The influence of the lateral front on the development of the windsea can be studied in more detail by considering the steady-state total energies E_{IV} at the two grid points $A \equiv (x^*, y^*) = (0.34 \times 10^7, 0.4 \times 10^7)$ and $B \equiv (x^*, y^*) = (0.34 \times 10^7, 1.0 \times 10^7)$ (cf. Table 9.1 and Fig. 9.3). To separate the effects of the wind field geometry from the different intrinsic growth rates of

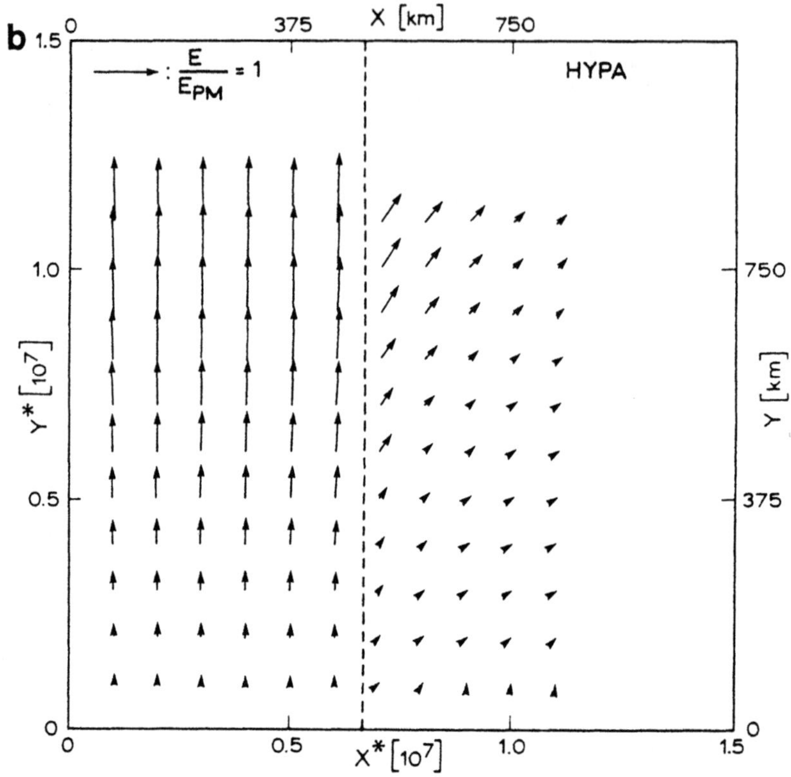

Fig. 9.2. (*continued*)

the models, the energies have been normalized in Table 9.1 and Fig. 9.3 by the corresponding fetch-limited total energies E_{II} for Case II. The grid points were chosen along the center line of the windy half-plane to minimize effects of directional asymmetry.

The differences in the ratios $E_{\mathrm{IV}}/E_{\mathrm{II}}$ can be explained by the form of directional coupling and the angular distributions of the windsea and input source functions, in close analogy to the growth rates along the symmetry diagonal discussed in Case III. The hybrid models exhibit very few differences relative to Case II, the observed small growth retardations arising mainly through numerical lateral diffusion in the advection schemes. SAIL again experiences the strongest growth reduction for the point A closest to the shore. In general, the strongest lateral boundary effects are found in the discrete models.

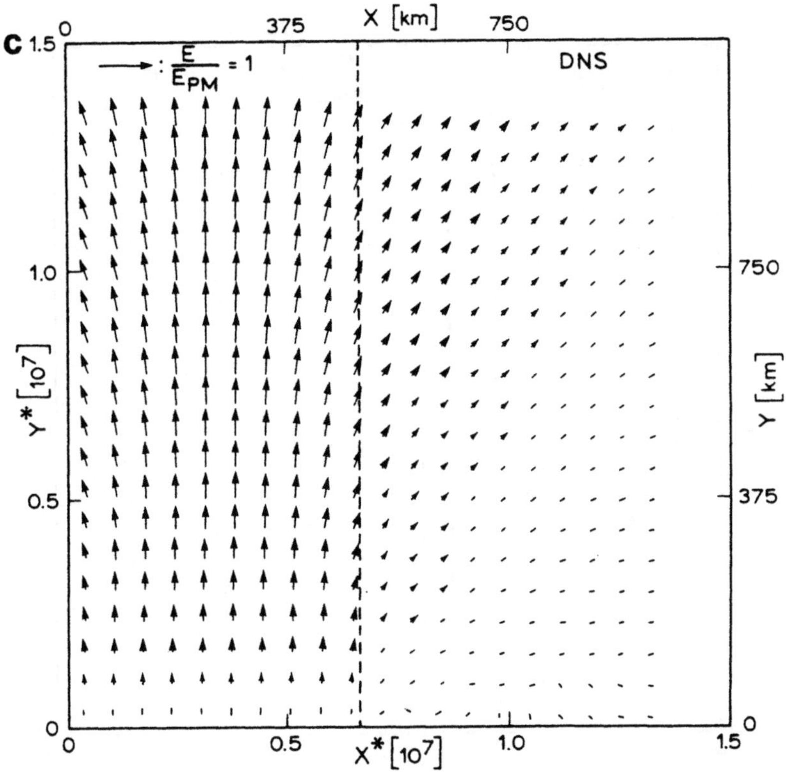

Fig. 9.2. *(continued)*

9.5. THE SWELL FIELD IN THE CALM HALF-PLANE

A more detailed insight into the structure of the swell field represented by the Custer diagrams, Fig. 9.2, is provided by the 2d spectral distributions of Fig. 9.4 (with another selection of model examples) for the point $C \equiv (x^*, y^*) = (1.01 \times 10^7, 1.01 \times 10^7)$ in the calm half-plane. The swell spectra clearly reflect the windsea directional distributions from which the swell components have radiated (broad distributions for MRI, panel b, and GONO, panel c, a narrower distribution for VENICE, panel a). The SAIL spectrum (panel d) exhibits an interesting skew away from the wind at low frequencies which is not present in the other models and appears to be due to the zero-lateral-gradient boundary condition which was applied at the left boundary ($x = 0$), which ensures an effective infinite fetch for laterally traveling waves.

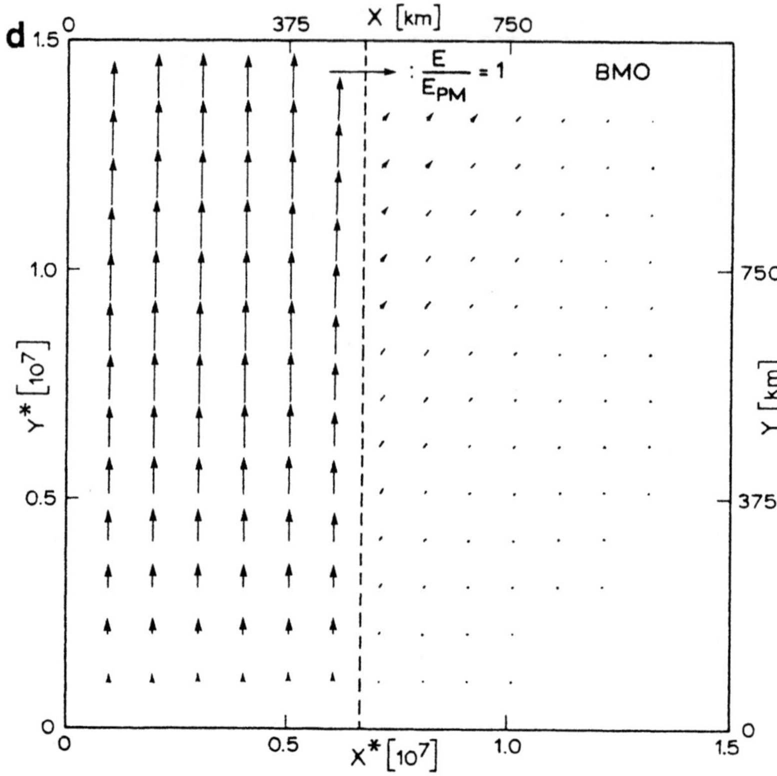

Fig. 9.2. (*continued*)

Table 9.2 lists the peak frequencies and mean swell directions for all models at the point C. The energies and frequencies have been normalized by the corresponding values at the point B at the same distance offshore in the windy half-plane (cf. Fig. 9.1). The data of Table 9.2 are summarized graphically in the parameter plane E_C/E_B versus $\bar{f}_C/\bar{f}_B$ in Fig. 9.5. (Arrows indicate the mean swell direction at C.)

As in the windy half-plane, the differences in the swell predictions do not group readily into model classes, but reflect primarily individual features of the growth rates and angular distributions of the wind seas from which the swells have radiated.

With the exception of the models MRI and NOWAMO, the swell–windsea energy ratios E_C/E_B (Fig. 9.5) are comparable for all models. The energy ratio is relatively high for the DP model MRI because of the rather broad directional distribution of the growth coefficients A and B in this model (cf.

Table 9.1
Ratios of the Windsea Energies in Case IV to the Pure Fetch-Limited Development in Case II for the Points A, B[a]

	A: $X^* = 0.34 \times 10^7$, $Y^* = 0.40 \times 10^7$			B: $X^* = 0.34 \times 10^7$, $Y^* = 1.01 \times 10^7$		
Model	E_{IV}/E_{PM}	E_{II}/E_{PM}	E_{IV}/E_{II}	E_{IV}/E_{PM}	E_{II}/E_{PM}	E_{IV}/E_{II}
MRI	0.39	0.42	0.93	0.51	0.58	0.88
VENICE	0.45	0.54	0.83	0.66	0.76	0.87
NOWAMO	0.37	0.38	0.97	1.00	1.21	0.83
GONO	0.47	0.47	1.00	0.67	0.68	0.99
TOHOKU	0.35	0.35	1.00	0.63	0.63	1.00
HYPA	0.39	0.39	1.00	0.98	0.98	1.00
BMO	0.46	0.49	0.94	0.90	0.92	0.98
SAIL	0.42	0.76	0.55	0.99	1.00	0.99
DNS	0.37	0.43	0.86	0.57	0.73	0.78

[a] Cf. Figs. 9.1 and 9.3

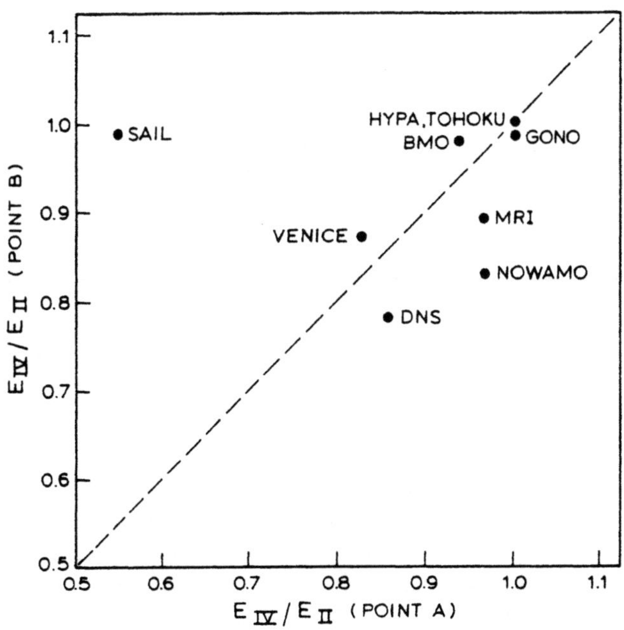

Fig. 9.3. Model locations in the parameter plane spanned by the values of (E_{IV}/E_{II}) at points A and B (Case IV, Fig. 9.1). In general, lateral boundary effects are weak, with somewhat stronger effects appearing in the discrete models.

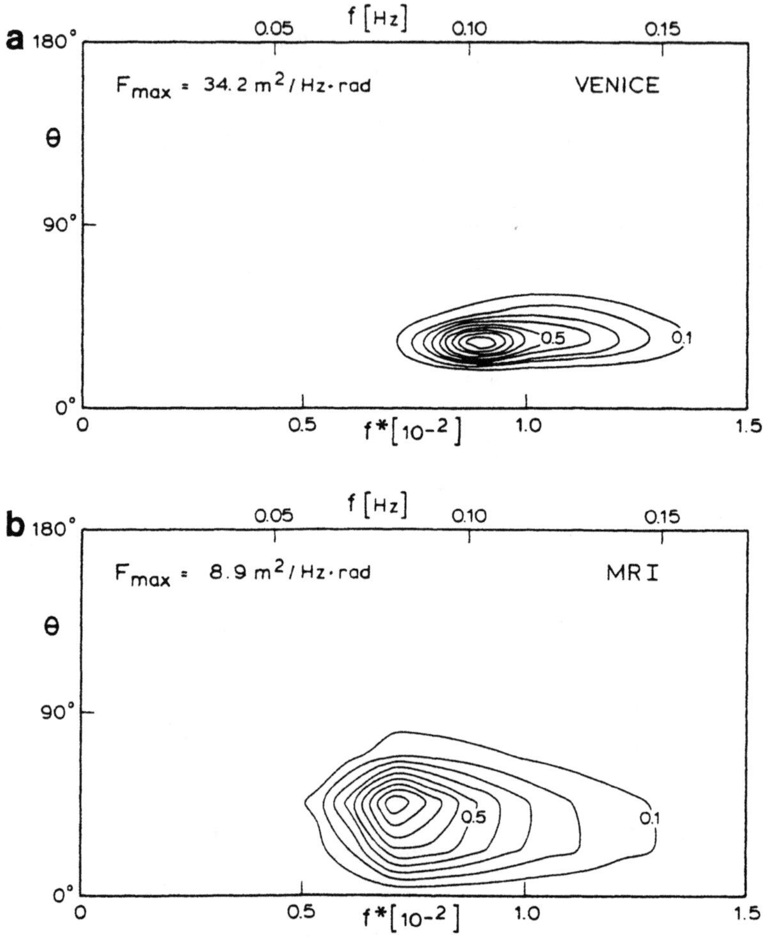

Fig. 9.4. Examples of 2d spectral distributions at point C (Case IV, Fig.9.1) for the DP models VENICE (panel a) and MRI (panel b), the CH model GONO (panel c), and the CD model SAIL (panel d). Contours are shown at intervals of 0.1 and in units of F/F_{max}.

Table 4.1). This yields relatively long fetches for the independent growth of wave components traveling at an angle to the wind, which appear later as swell at point C. The exceptionally low energy ratio for NOWAMO is due to the fact that this model retains only one directional bin in the swell output spectrum per frequency band.

Relatively few variations are seen in Fig. 9.5 in the frequency ratios and mean directions. These quantities depend on the relative contribution of the swell components propagating at a small angle to the wind from less developed windsea regions near the coast as compared with components

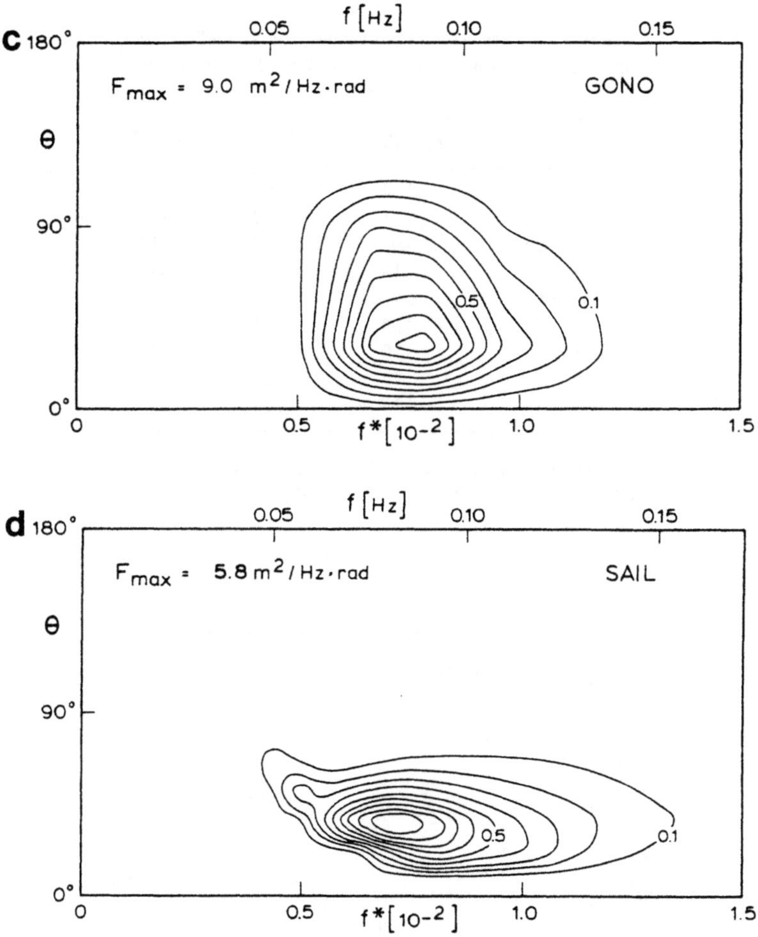

Fig. 9.4. (*continued*)

propagating at larger angles to the wind from further developed windseas at larger fetches. The net effect depends on the competing factors of growth rate and directional spread, which are difficult to classify in general terms without entering into model details.

9.6. CONCLUSIONS AND RECOMMENDATIONS

The effect of a lateral wind field boundary on wave growth in the interior of the wind field is qualitatively similar to the effect of the angled boundaries on wave growth along the diagonal which was discussed in the

Table 9.2
Relative Energies, Mean Frequencies, and Mean Directions of Swell in the Calm Half-Plane at $X^* = Y^* = 1.01 \times 10^7$, Point C[a]

Model	E_C/E_B	$\bar{f}_C/\bar{f}_B$	$\bar{\theta}_C - \bar{\theta}_B$
MRI	0.24	1.05	39°
VENICE	0.11	1.19	30°
NOWAMO	< 0.05	1.18	45°
GONO	0.14	0.97	47°
TOHOKU	0	—	—
HYPA	0.18	1.24	45°
BMO	0.10	1.22	47°
SAIL	0.13	1.26	35°
DNS	0.17	1.06	45°

[a] The quantities in the windy half-plane (subscript B) are taken at $(X^*, Y^*) = (0.34 \times 10^7, 1.01 \times 10^7)$, point B.

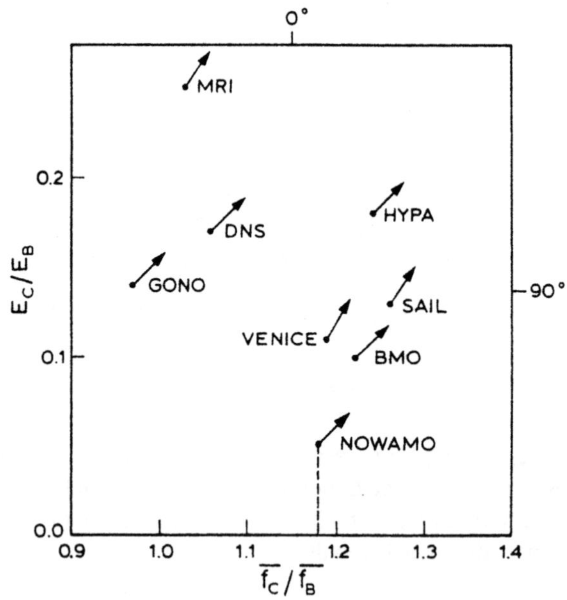

Fig. 9.5. Model location in parameter plane of E_C/E_B versus $\bar{f}_C/\bar{f}_B$ for Case IV. Indices B, C refer to points B, C (Fig. 9.1); $\bar{f}$ denotes the mean frequency averaged over the spectrum. Arrows indicate the mean swell directions at point C.

slanting fetch Case III. Waves propagating at an angle to the wind are exposed to less fetch than in the standard fetch-limited Case II. This affects discrete spectral models, particular those with broad spreading functions, more strongly than hybrid models, which do not follow the evolution of individual windsea components.

The swell radiation into the calm half-plane could be simulated reasonably well by all models (with the exception of a specific numerical problem encountered by the TOHOKU model for this particular geometry, which has since been resolved). Significant deviations in the swell–windsea energy ratios were found for two models (MRI and NOWAMO) relative to the model average. These could be explained by the relatively broad windsea directional distribution of MRI and the single-directional bin swell output algorithm of NOWAMO.

Hybrid models are unable to model a completely continuous spectral transition across a wind field discontinuity. Apart from this handicap, however, model performances were found to be comparable for all three model classes.

Most of the features of the swell field observed in this test could be explained directly by the directional properties of the windsea which radiated the swell into the calm half plane. Apart from this obvious model dependency, the test case revealed no particularly critical elements in any of the model types. Experimentally, a test of the model directional properties is probably easier to carry out under field conditions corresponding to Case III than the present case.

Chapter 10

Case VII: 90° Change in Wind Direction

10.1. WIND FIELD, BOUNDARY CONDITIONS

A uniform wind blows to the north at 20 m/s over an infinite ocean until the sea is fully developed ($f_p = f_{PM}$, Case VIIa) or half-developed ($f_p = 2f_{PM}$, Case VIIb). At this time ($t = 0$) the wind is instantly turned 90°, from north to west, at the same wind speed, 20 m/s. As the wind and wave fields are homogeneous in space, the wave models can be run, without the advection terms, at a single "grid point" as a function of time only. (Where this could not be readily implemented, the asymptotically homogeneous downwind limit of a space–time-dependent solution was taken, in analogy to the duration-limited results of Case II.)

10.2. PURPOSE OF THE TEST

At the instant the wind direction turns, the preexisting windsea largely becomes "swell," and a new windsea begins to grow. The time evolution of the swell and windsea results in a relaxation of the wave field toward a new steady sea state which approaches, asymptotically, a fully developed windsea propagating in the new wind direction. The purpose of the test was to study this relaxation process. The test may also be seen as a simpler time-dependent analogue of the inhomogeneous, stationary front considered in Case V (cf. Chapter 11).

10.3. RELAXATION MECHANISMS

In general, three mechanisms contribute to the directional relaxation: (1) energy input by the wind to the new windsea, (2) attenuation of the swell, and (3) weak nonlinear interactions, which transfer energy from the swell to the windsea.

The different simulations of these processes by different wave models yield significant deviations of the results for this case. It should be noted, however, that because of the large directional change, the test exercises only those interaction processes which are related to swell–windsea coupling. Most models contain additional directional response mechanisms involving interactions within the windsea spectrum itself, which are important for a continuously turning wind vector, but which do not come to bear in the present situation.

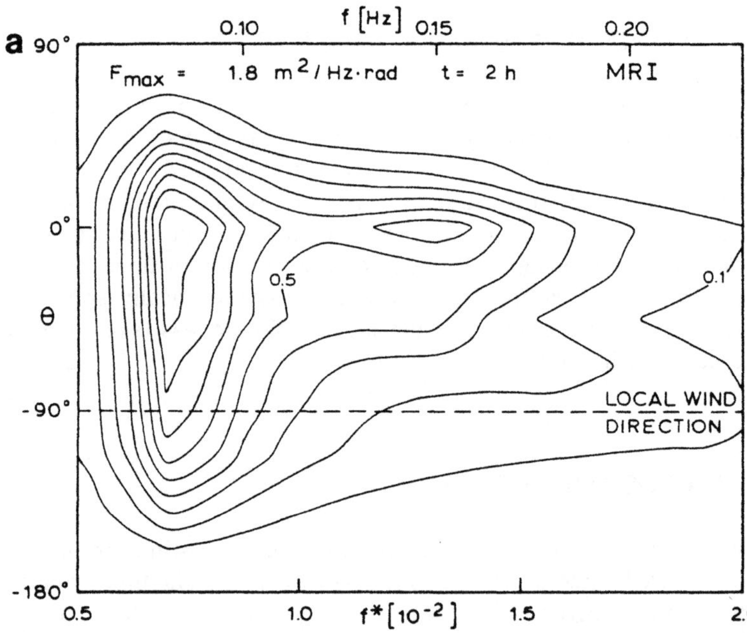

Fig. 10.1. Contour plots of the 2d spectral distributions at two characteristic times after a sudden change of wind direction from 0° to −90° for the DP model MRI (Case VII), Figures 10.2–10.10 display similar results for the other models. For this test, the spectra were half-developed ($f_p = 2f_{PM}$) at the time of the wind shift ($t = 0$), with the exception of the results for the CD model DNS (Fig. 10.9), for which a fully developed spectrum ($f_p = f_{PM}$) was

Of the three processes, the swell–windsea coupling is responsible for the largest deviations.

The first-generation DP models contain no coupling. For these models the test case simply describes the decay (if any) of the swell field produced by the switch in wind direction and the simultaneous, independent generation of a new, superimposed windsea.

Among the coupled models we may distinguish between models in which the transition from windsea to swell is described in terms of relative frequency relations (only HYPA) and models which incorporate in addition some form of dependence on the relative directional characteristics of windsea and swell components (GONO, NOWAMO, SAIL, TOHOKU, DNS, and BMO). In most models the swell–windsea transfer algorithm is energy-conserving, but in some models energy is lost in the transition process. The details of the coupling cannot be grouped readily into the classes CH or CD but depend rather on the individual model hypotheses. Exact computations of the nonlinear energy transfer for a number of superimposed swell and windsea spectra (Hasselmann, 1963b) indicate that the coupling is weak until the swell and windsea frequencies become comparable, and then

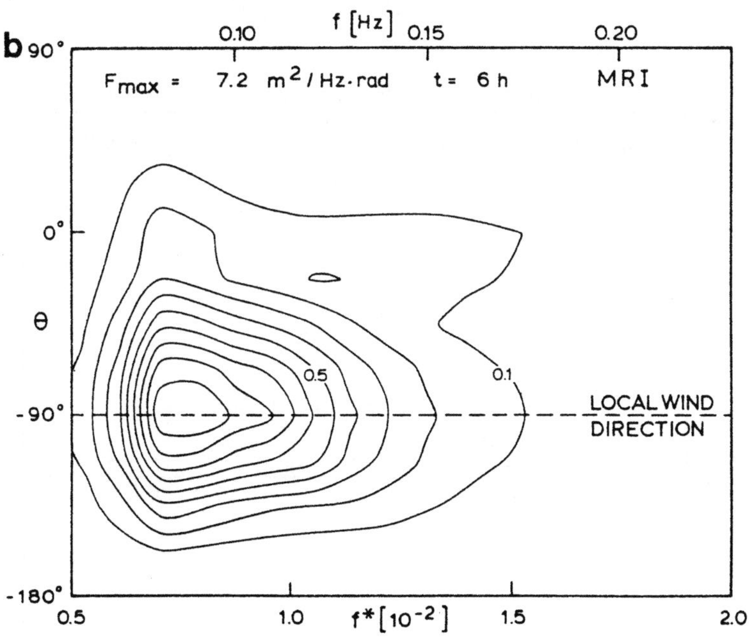

assumed at $t = 0$. Each spectrum is normalized by its maximum value, and contours are shown at intervals of 0.1 (note that this can give rise to an apparent decrease of the swell peak when the windsea peak energy density increases beyond the swell peak density).

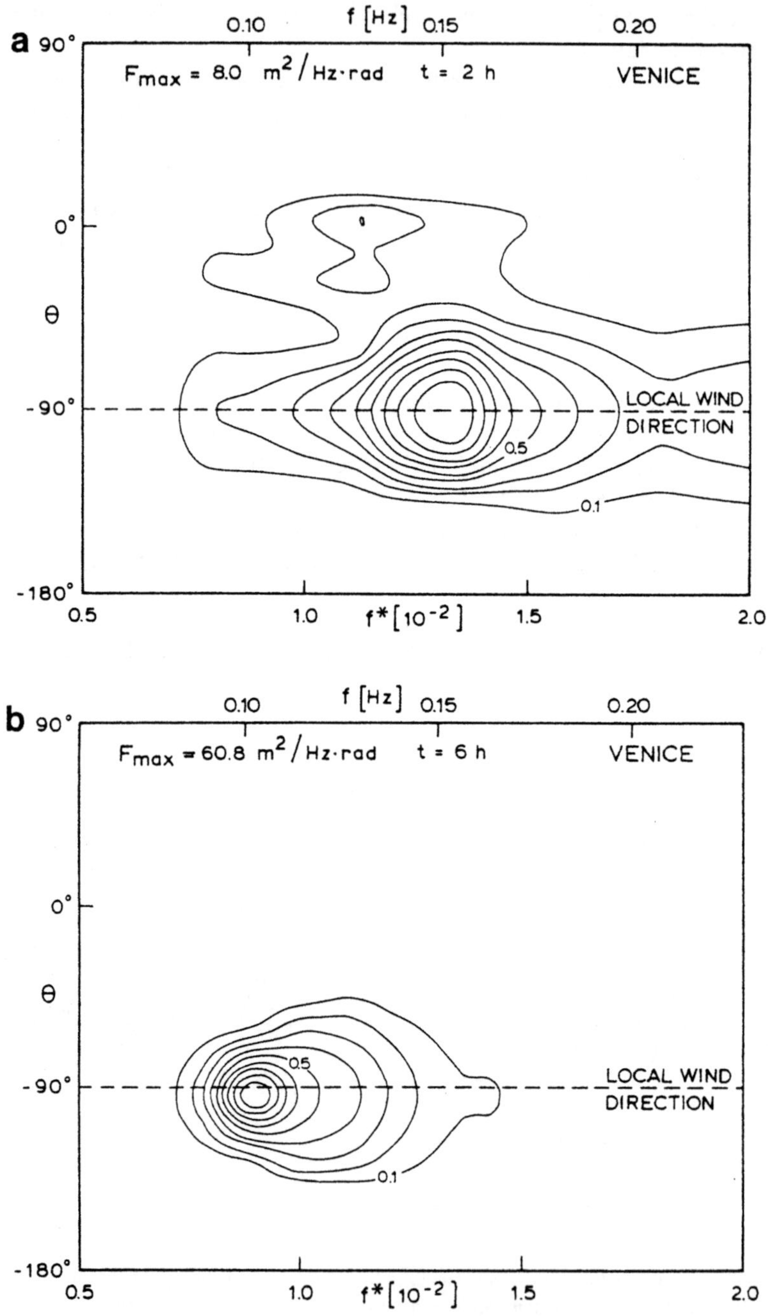

Fig. 10.2. Same as Fig. 10.1 for the DP model VENICE (Case VII).

rapidly increases. The transition region is rather insensitive to the relative propagation directions of windsea and swell. This is supported by the EXACT-NL computation for this case.

10.4. RESULTS

The principal results of this test case are summarized in the 2d spectral distributions of Figs. 10.1–10.10. The spectra are shown at, typically, 2 and 6 h after the wind direction has turned, and for all models except DNS the spectra were half-developed at the time of the wind shift ($f_p = 2f_{PM}$). Only the fully developed initial state case ($f_p = f_{PM}$) was run on the DNS model, and corresponding results at $t = 1$ and 30 h are displayed. For the NOWAMO and GONO models, output data were available only at $t = 6$ and 7.5 h, respectively. For the models on which both cases were run, qualitatively similar results were obtained for both half-developed and fully developed initial states (cf. complete compilation of results in SWAMP group, 1982).

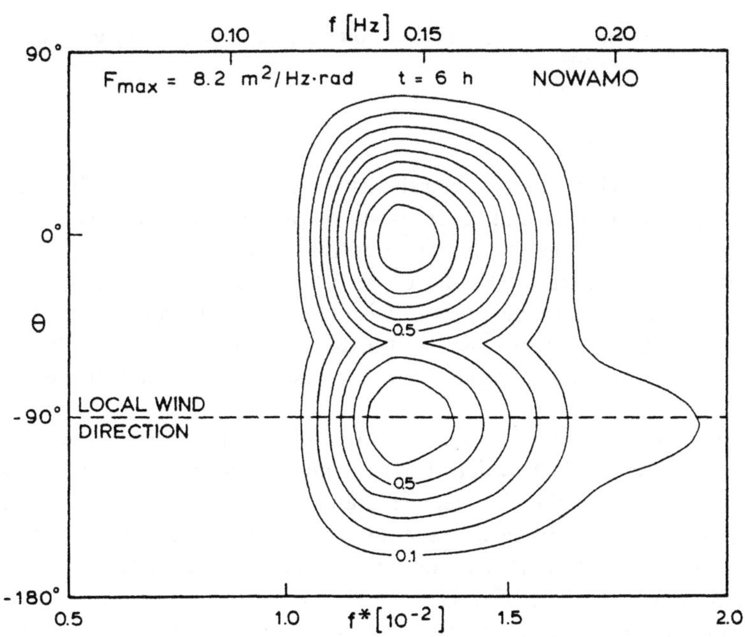

Fig. 10.3. Same as Fig. 10.1 for the CH model NOWAMO; for this model, no output was available at less than 6 h (Case VII).

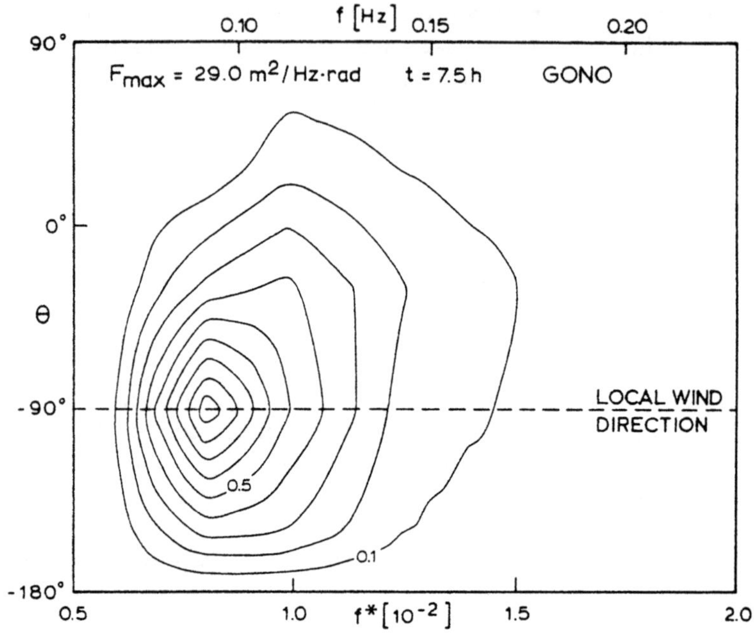

Fig. 10.4. Same as Fig. 10.1 for the CH model GONO; for this model, no output was available at less than 7.5 h (Case VII).

The wide variety of 2d spectral distributions seen in Figs. 10.1–10.10 largely reflects the fact that there exists at present no generally accepted parameterization of the nonlinear coupling between windsea and swell. Thus all models (except EXACT-NL) introduce rather arbitrary, intuitive assumptions to simulate this process, with only weak guidance from existing exact computations for special spectral distributions. However, differences in swell dissipation and windsea growth rates also contribute to the deviations.

The EXACT-NL computations (Fig. 10.10) indicate that for the present case of a 90° wind shift, which leads to the generation of a new windsea with initially much higher frequencies than the former windsea (new swell), the windsea–swell coupling is essentially negligible until the new windsea peak frequency has decreased almost to the value of the swell peak frequency (at $t = 6$ h, cf. 10.10b). The growth of the new windsea at $\theta = 90°$ is practically identical to the normal duration-limited growth case without swell (Case II), while the swell decay is controlled primarily by the dissipation. (This depends, however, for the quasi-linear dissipation expression of

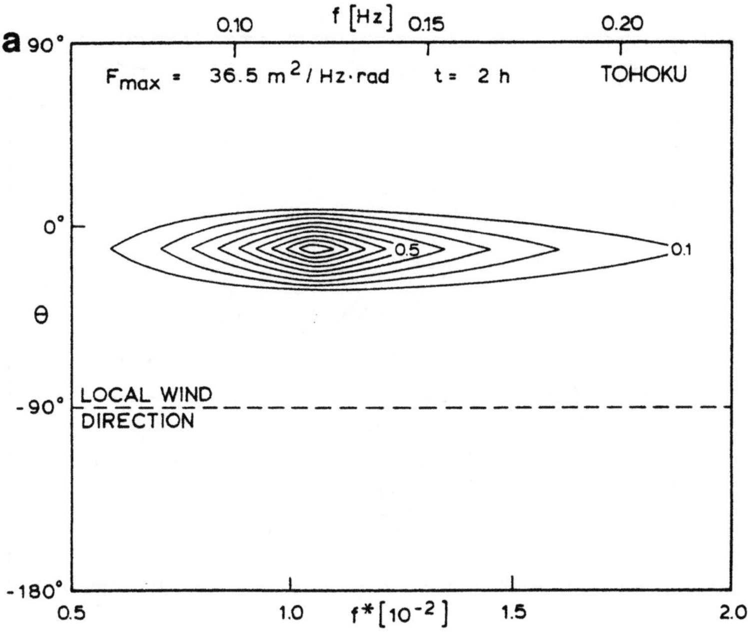

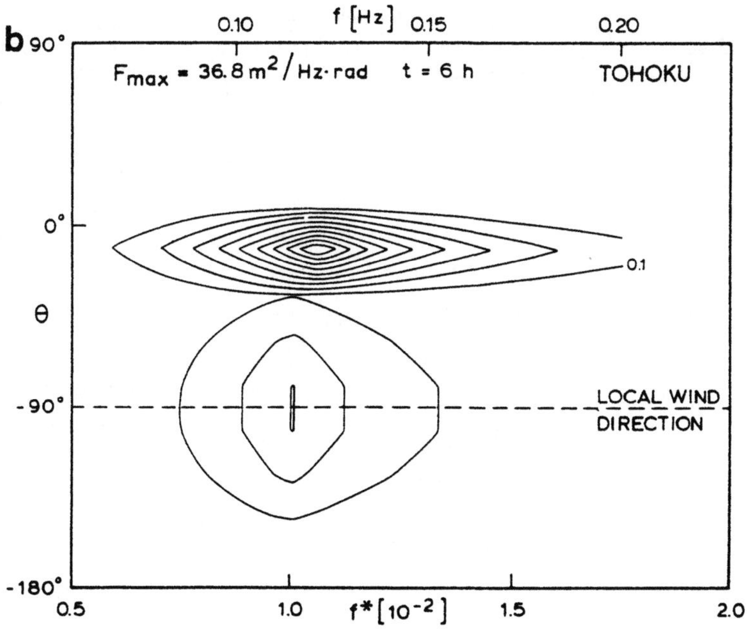

Fig. 10.5. Same as Fig. 10.1 for the CH model TOHOKU (Case VII).

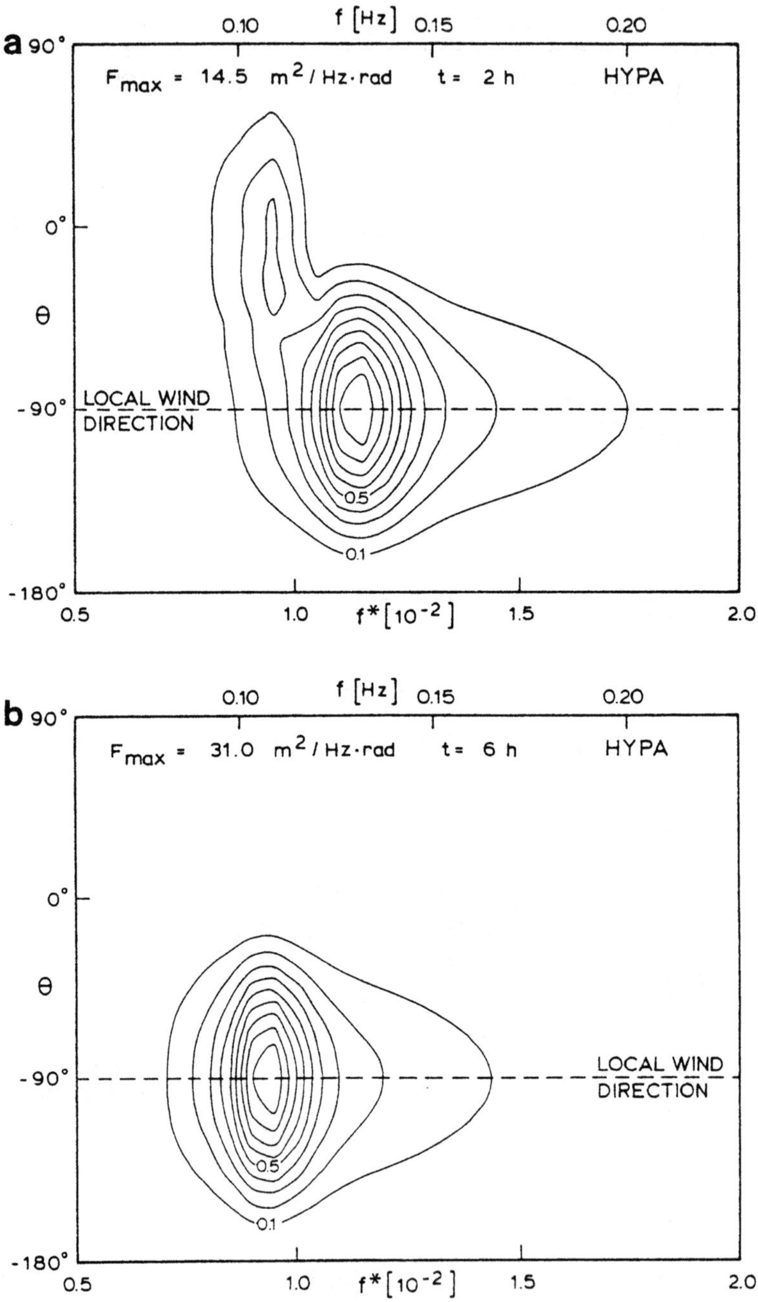

Fig. 10.6. Same as Fig. 10.1 for the CH model HYPA (Case VII).

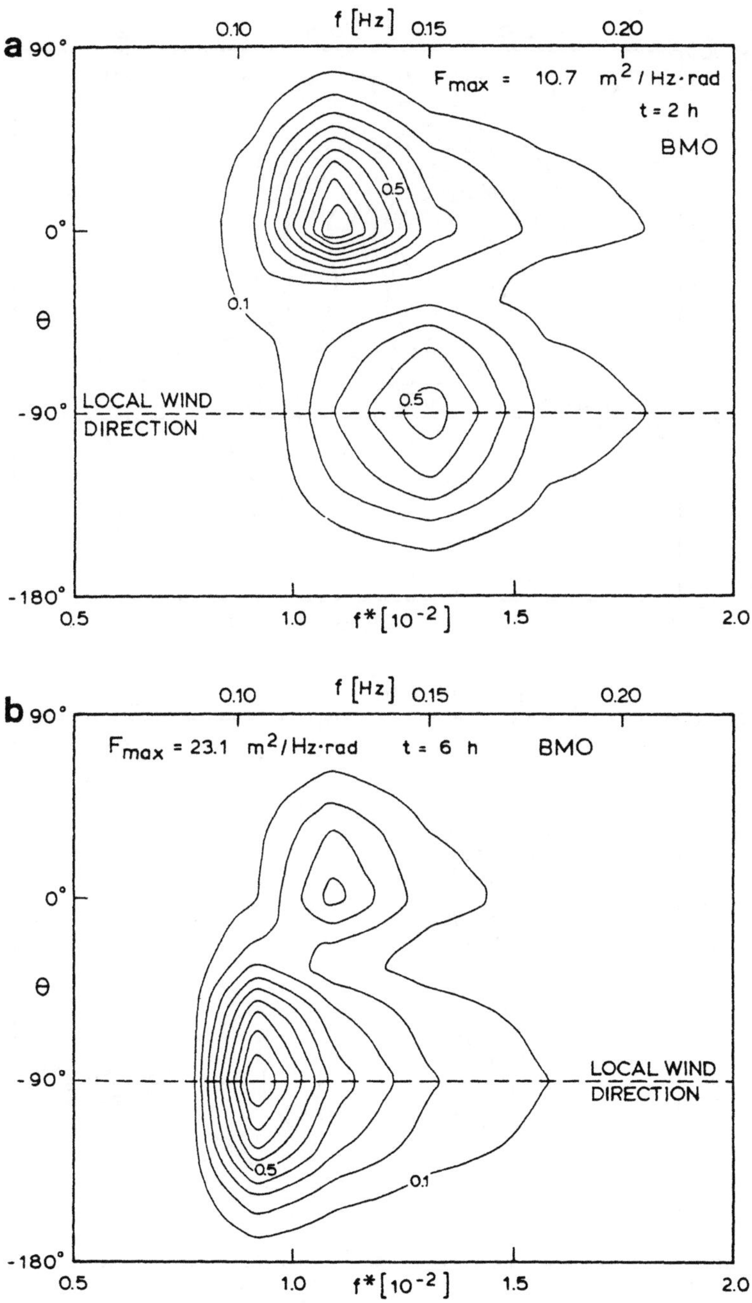

Fig. 10.7. Same as Fig. 10.1 for the CD model BMO (Case VII).

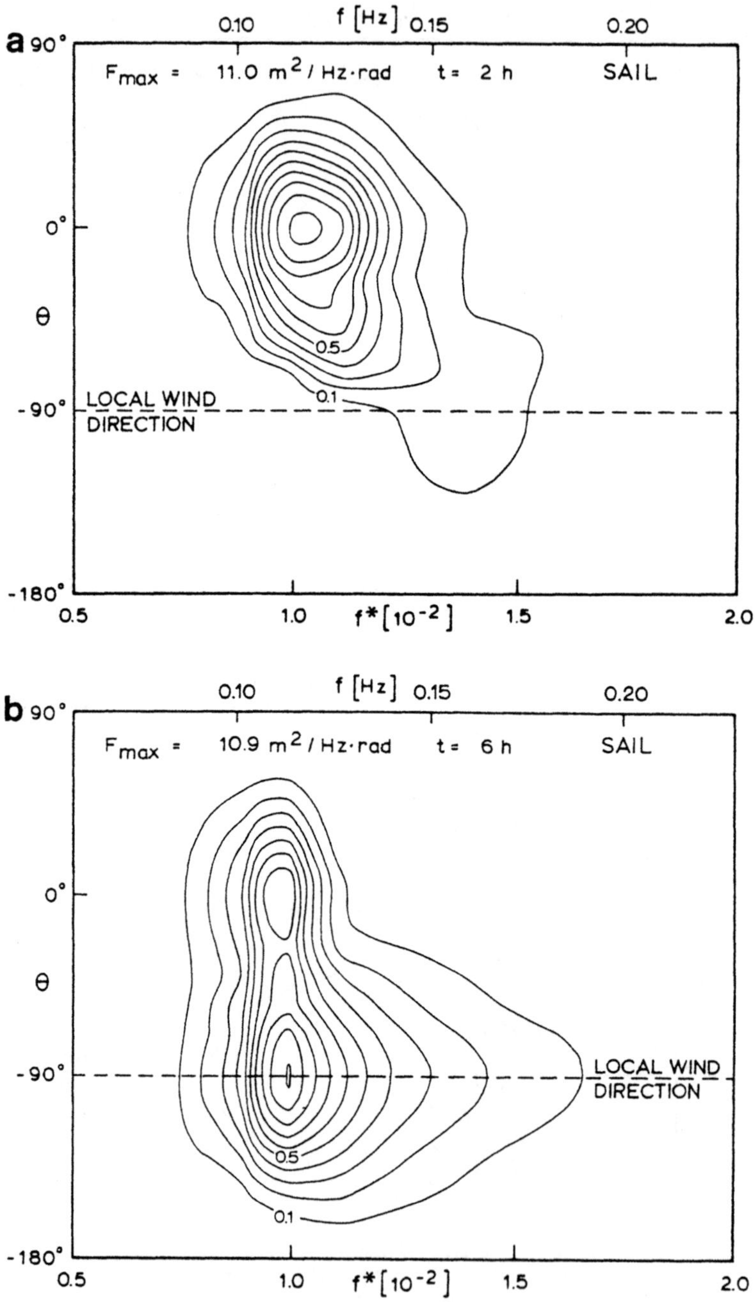

Fig. 10.8. Same as Fig. 10.1 for the CD model SAIL (Case VII).

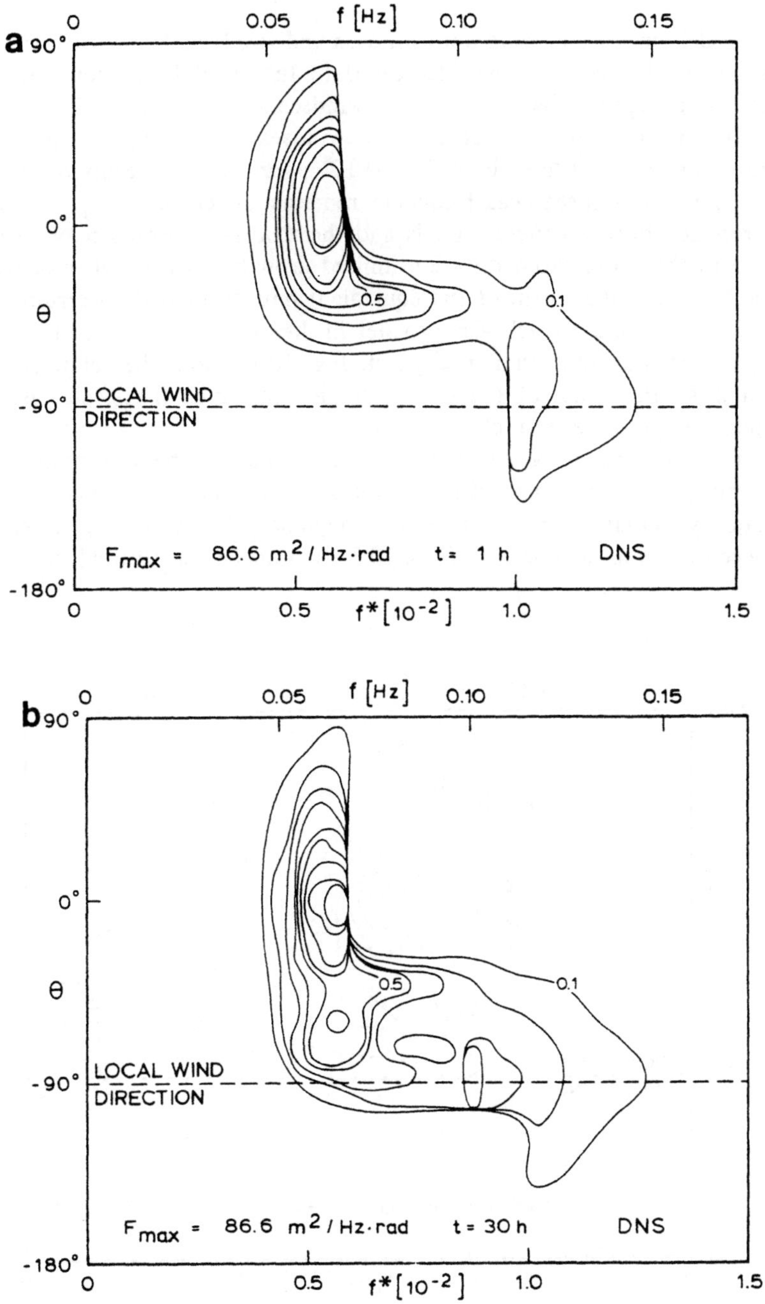

Fig. 10.9. Same as Fig. 10.1 for the CD model DNS, except that a fully developed spectrum was assumed at the time the wind turned. Output was available only at $t = 1$ and $t = 30$ h for this model (Case VII).

this model, on the energy in both windsea and swell, cf. Table 6.1.) After the windsea peak frequency has decreased to the swell frequency, the swell spectrum is rapidly absorbed into the windsea spectrum (Fig. 10.10c).

It appears that the neglect of all windsea–swell coupling in the DP models MRI and VENICE (Figs. 10.1–10.2) is an acceptable approximation while the new windsea peak frequency remains higher than $2f_{PM}$. The main discrepancy between these models and the EXACT-NL results may be attributed to the more rapid duration-limited growth rates of DP models (cf. Case II), the initial form of the equivalent half-developed spectrum taken for the MRI model (which was broader in direction and extended to lower frequencies than the other models, cf. Fig. 10.1a), and the relatively high swell dissipation rate of the VENICE model (in which the swell has completely decayed after 6 h, cf. Fig. 10.2b).

For the case of a 90° wind direction change, the CH models also incorporate no windsea–swell coupling while the new windsea peak frequency remains above the swell frequency. Most of the differences between these models seen in Figs. 10.3–10.6 may be attributed to the way

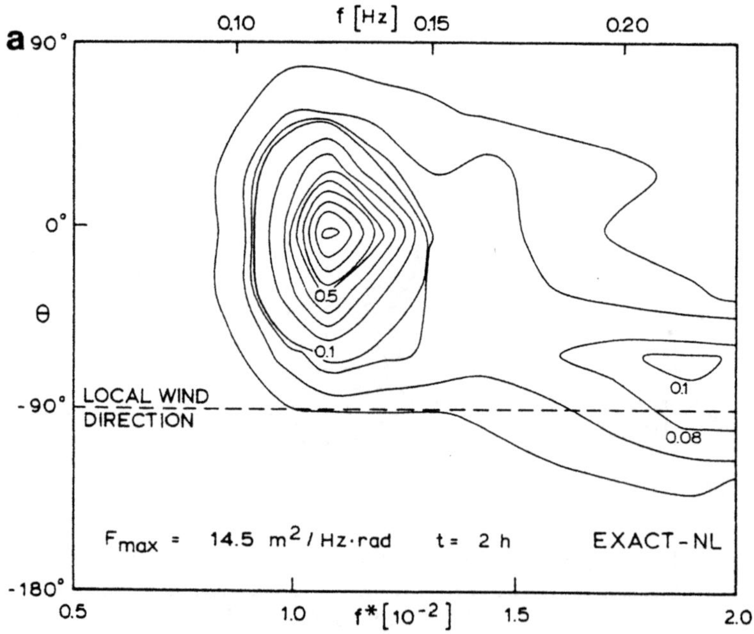

Fig. 10.10. Same as Fig. 10.1 at three output times for the EXACT-NL model. Three additional contours have been inserted at 0.01, 0.04, and 0.08 (Case VII).

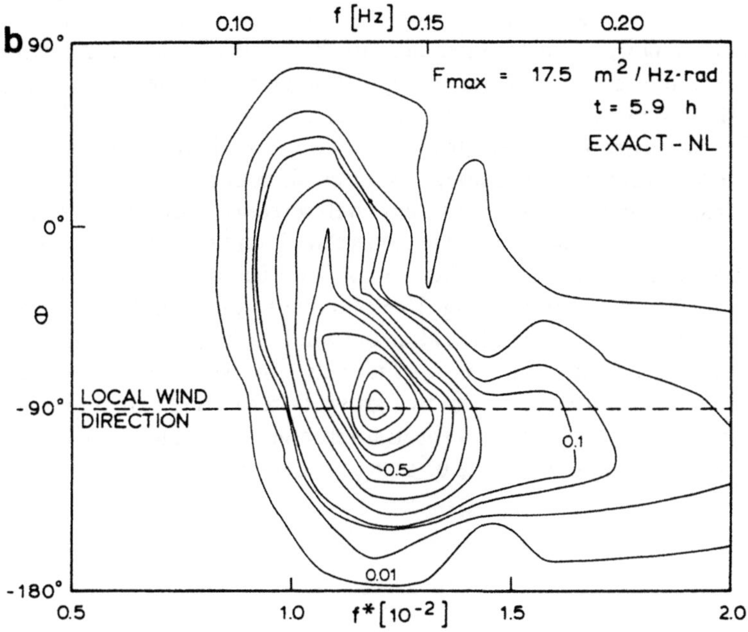

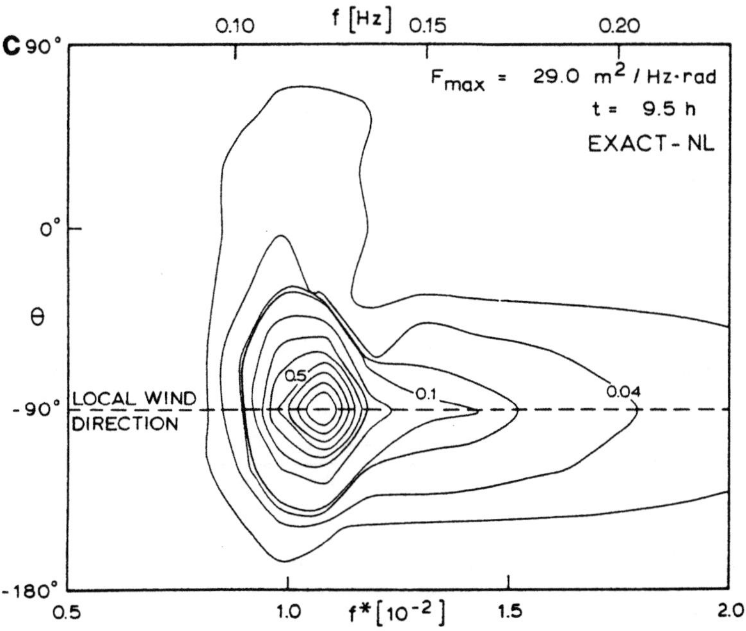

Fig. 10.10. (*continued*)

in which the windsea is transferred into swell at the instant the wind direction changes and to differences in the swell dissipation rate. TOHOKU transfers the entire windsea energy at a given frequency into a single direction, NOWAMO retains only the directional bin with the highest energy (which is subsequently redistributed as a $\cos^2\theta$ distribution), while GONO and HYPA transfer the assumed windsea directional distribution without significant change into an equivalent swell directional distribution. TOHOKU, HYPA, and GONO include no or negligible swell damping, while NOWAMO has dissipated all swell energy after 18 h. The disappearance of swell in the HYPA spectrum within 6 h (Fig. 10.6b) is due to the complete absorption of the swell into the new windsea after the windsea frequency has decreased to the swell frequency. This is qualitatively consistent with the EXACT-NL results except that the new windsea grows faster in HYPA than in EXACT-NL.

Because of their uniform spectral representation, the CD models (Figs. 10.7–10.9) should in principle be able to simulate a smoother windsea–swell

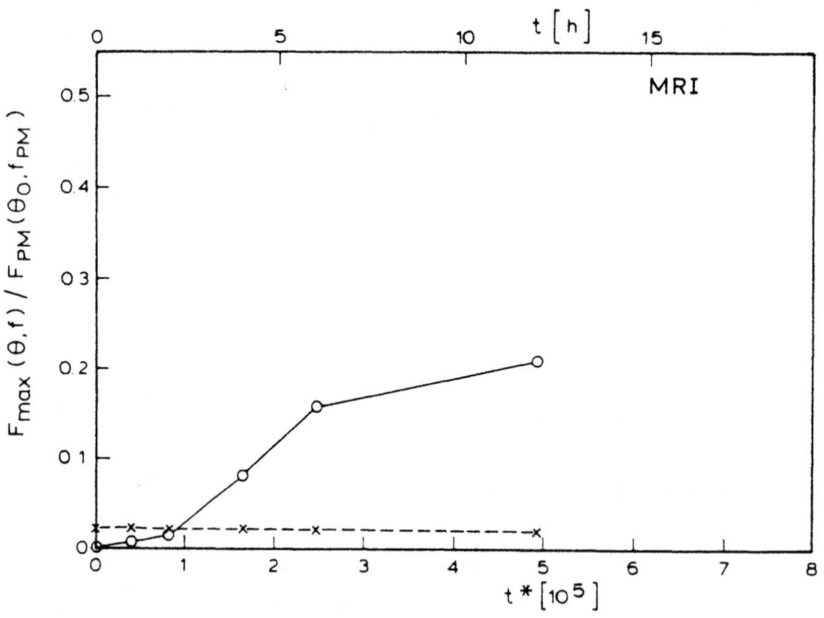

Fig. 10.11. Evolutions of the windsea (solid line) and swell (dashed line) peak spectral densities for the DP model MRI after the change in wind direction at $t = 0$ for an initially half-developed windsea. Figures 10.12–10.19 show corresponding curves for other participating models (no figure is shown for DNS, for which insufficient output data were available) (Case VII).

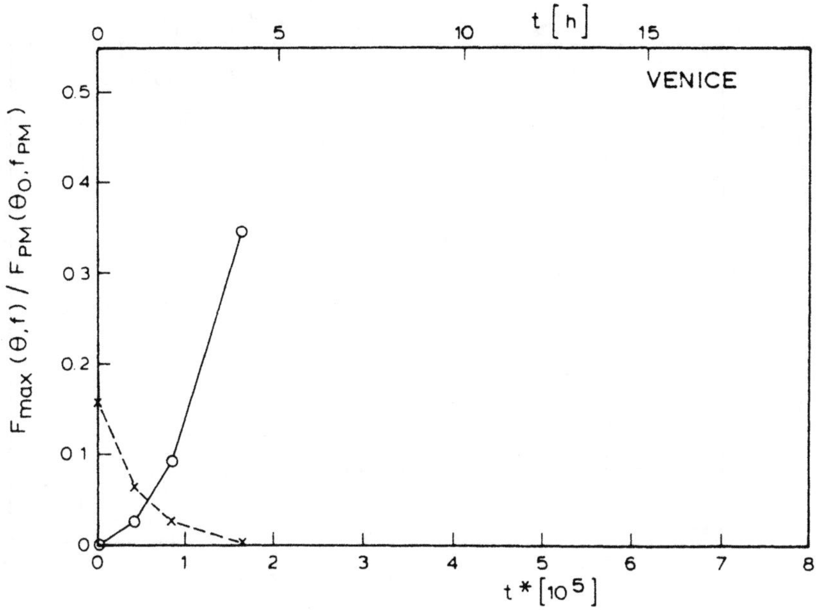

Fig. 10.12. Same as Fig. 10.11 for the DP model VENICE (Case VII).

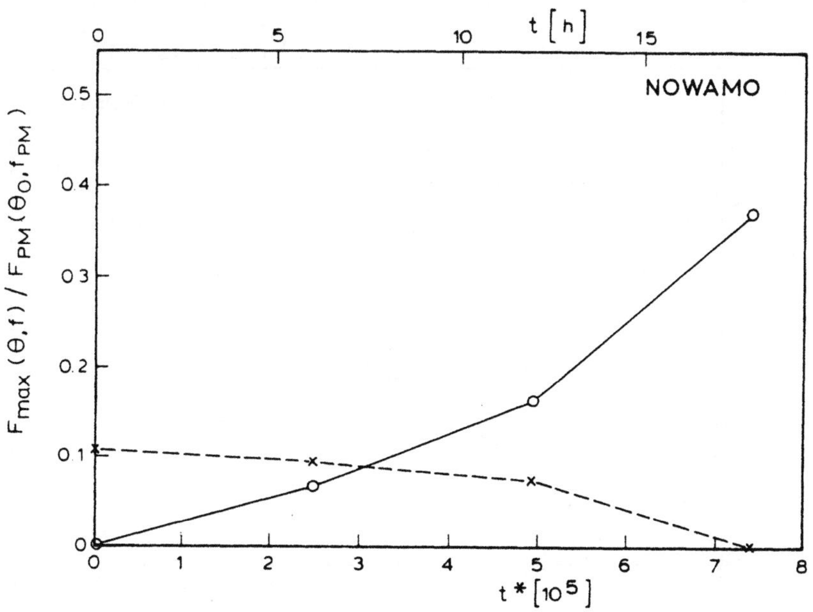

Fig. 10.13. Same as Fig. 10.11 for the CH model NOWAMO (Case VII).

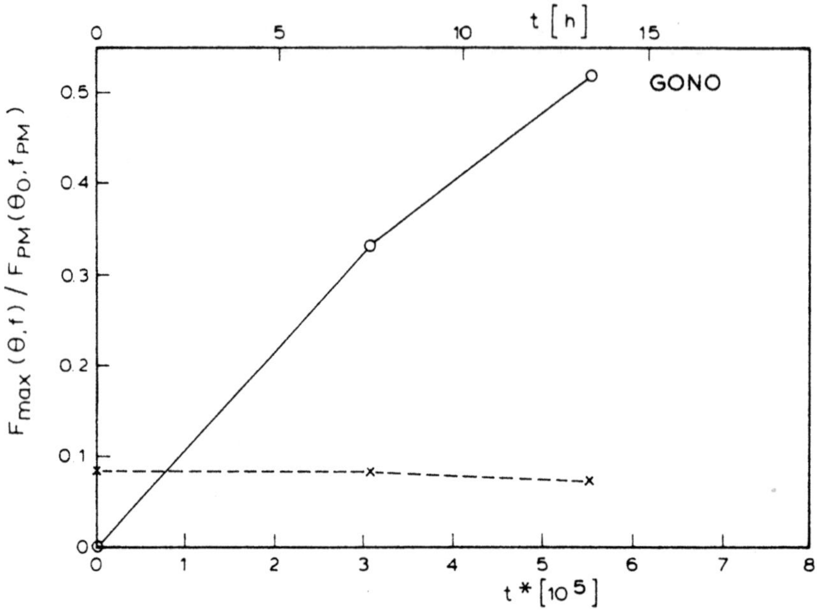

Fig. 10.14. Same as Fig. 10.11 for the CH model GONO (Case VII).

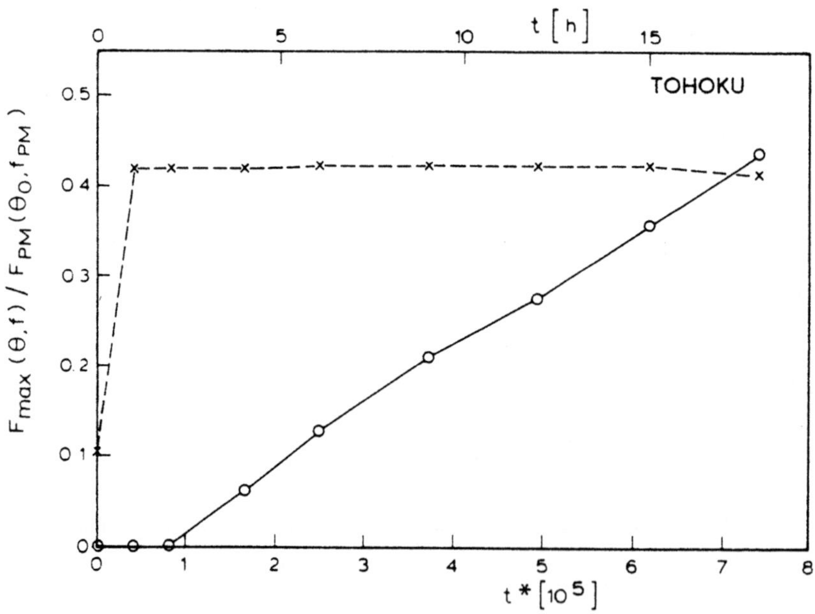

Fig. 10.15. Same as Fig. 10.11 for the CH model TOHOKU (Case VII).

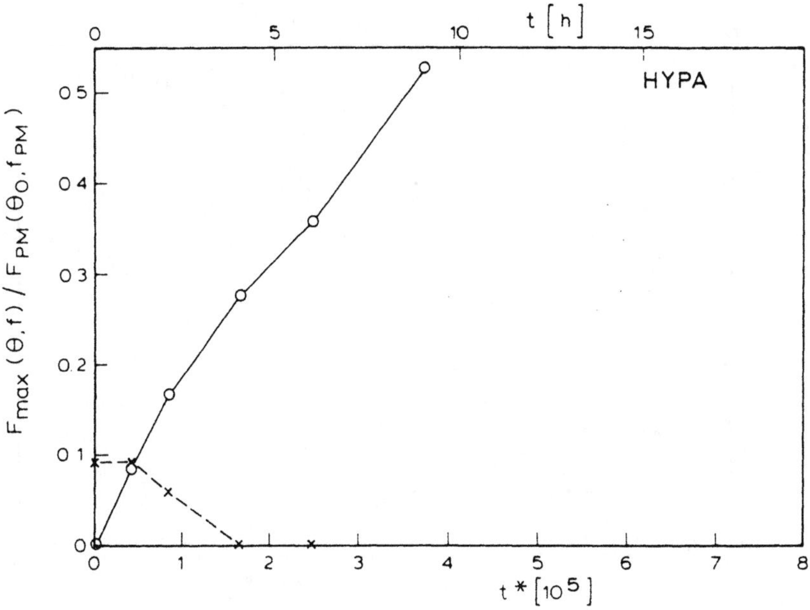

Fig. 10.16. Same as Fig. 10.11 for the CH model HYPA (Case VII).

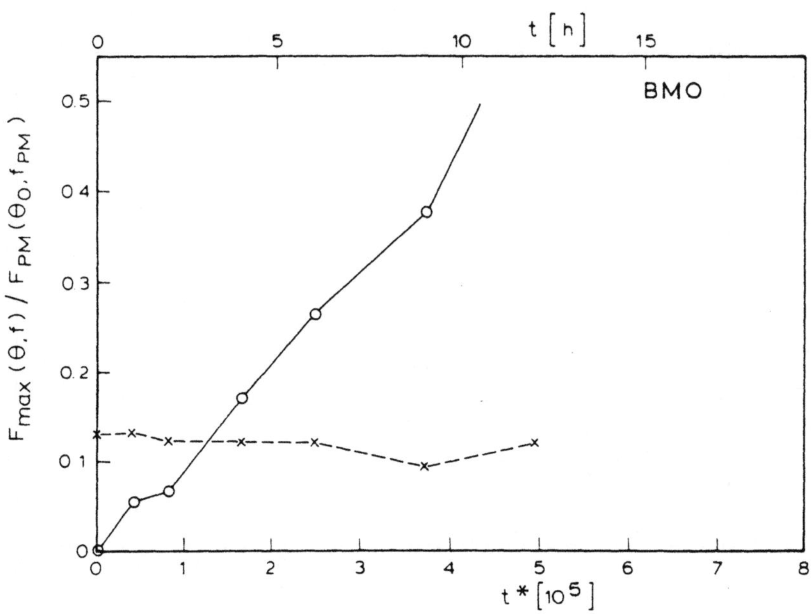

Fig. 10.17. Same as Fig. 10.11 for the CD model BMO (Case VII).

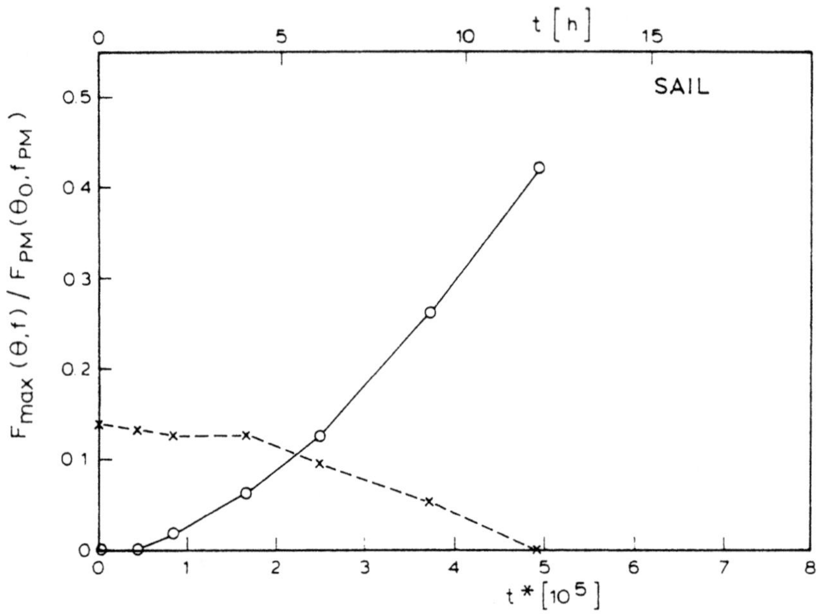

Fig. 10.18. Same as Fig. 10.11 for the CD model SAIL (Case VII).

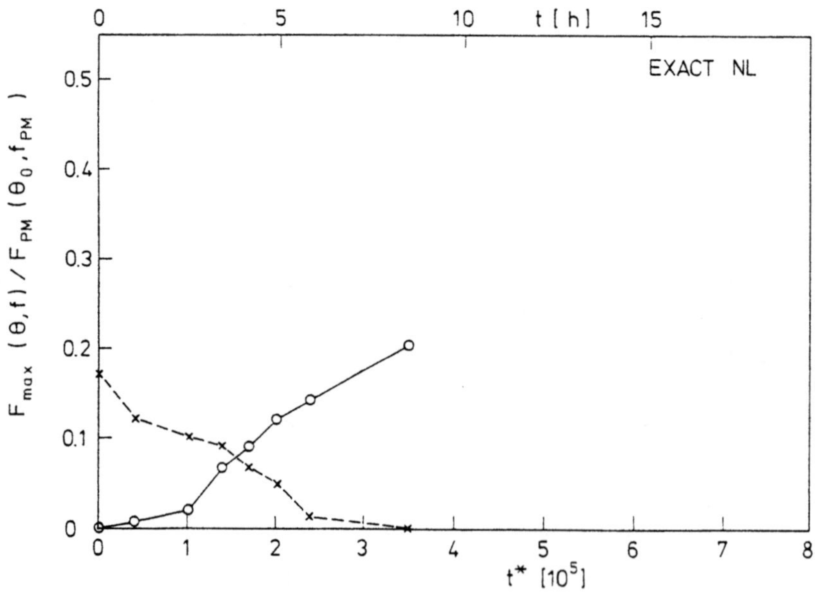

Fig. 10.19. Same as Fig. 10.11 for the model EXACT-NL (Case VII).

transition. However, with the exception of SAIL (Figs. 10.8a, b), which comes closest to the EXACT-NL simulation, the nonlinear windsea–swell interaction is also treated as a discontinuous process in these models. In the BMO model, swell components are absorbed into the windsea for frequencies greater than 0.8 of the windsea peak frequency and directions less than 90° from the windsea direction. The unusual DNS spectral distributions (Figs. 10.9a, b) result from a discontinuous windsea–swell coupling algorithm in which all swell components with frequencies greater than the Pierson–Moskowitz frequency are attenuated to conform to the limiting spectrum centered on the new wind direction. This annihilates wave components propagating at angles greater than 90° to the wind, while the surviving swell components with frequencies lower than the Pierson–Moskowitz frequency propagate undamped.

The relative growth of the windsea and decay of the swell are seen more clearly in the nonnormalized growth curves of Figs. 10.11–10.19. These show the relative maxima of the 2d spectral densities for the windsea and swell regions of the spectra (the DNS curves could not be constructed because of insufficient output data).

For all models the development of the new windsea was essentially identical to the normal duration-limited growth case. The linear swell decay was negligible for most of the coupled models, while a dramatic reduction of swell energy occurred for these models when the coupling conditions for nonlinear windsea–swell interactions were achieved.

10.6. CONCLUSIONS AND RECOMMENDATIONS

The test case clearly separated the models into two classes: those incorporating no windsea–swell coupling (the two DP models and the TOHOKU model), and those in which the coupling was important for the initial windsea–swell transition and the later reabsorption of the swell into the newly developed windsea (all CH and CD models, with the exception of the CH model TOHOKU). The first class can be divided further into models with and without significant dissipation, while the second class (for which the linear swell dissipation is generally insignificant) groups into three subclasses according to the criteria governing the windsea–swell coupling (cf. Table 10.1).

Qualitatively, all models show a similar independent superposition of windsea and swell while the new windsea is still in the early stages of development, but models of different classes or subclasses begin to diverge strongly after the new windsea has developed to a stage where nonlinear windsea–swell coupling can come into play. The times when this occurs

Table 10.1
Model Classification for Case VII

Windsea–swell interaction	Subclass	Model	Model class
Decoupled	With swell dissipation	VENICE	DP
	Without significant swell dissipation	MRI	DP
		TOHOKU	CH
Coupled	Frequency criterion	HYPA	CH
		SAIL	CD
	Frequency–direction criterion	DNS	CD
		BMO	CD
		GONO	CH
		NOWAMO	CH
	Continuous coupling, more strongly frequency- than directionally dependent	EXACT-NL	CD

vary considerably, depending on the growth rates of the model (cf. Figs. 10.11–10.19).

The most critical process in this test case, the nonlinear coupling between windsea and swell, is basically understood. However, existing parameterizations of the nonlinear transfer have been derived from considerably simpler spectral distributions than those occurring in this test. The case study emphasizes the need for a general operator-type parameterization of the nonlinear transfer applicable for arbitrary spectral distributions.

In addition, the test revealed the wide range of swell damping rates assumed in present models. This is also indirectly related to differences in the modeling of the nonlinear absorption of swell by windsea.

Chapter 11

Case V: Diagonal Front

11.1. WIND FIELD, BOUNDARY CONDITIONS

A front runs diagonally from SW to NE across a square of ocean 1000 km on a side (Fig. 11.1). The initial condition is a calm sea, and boundary conditions are no influx of wave energy and perfect absorption at the boundaries. The wind blows to the north SE of the front and to the west NW of the front, in both sectors at the same speed $U_{10} = 20$ m/s. For grid points directly on the diagonal, the wind is taken to the north. The wave field is examined after the steady state has been achieved.

11.2. PURPOSE OF THE TEST

The original purpose of this case study was to determine the response of the model wave fields to turning winds. After the tests were carried out, it was decided to include also the simpler Case VII, discussed in the previous chapter. Although less realistic, Case VII isolates the directional relaxation effects, while the present case represents a more complex superposition of these effects with signatures associated with fetch-limited growth in the SE sector and oblique fetch generation (Case III) in the NW sector, after the waves have crossed the front.

Qualitatively, the following features may be expected: SE of the front, a fetch-limited wave field is established whose mean direction is essentially parallel to the wind (except where boundary effects become significant). Since the effective parallel fetch from the sheltering southern coastline to the

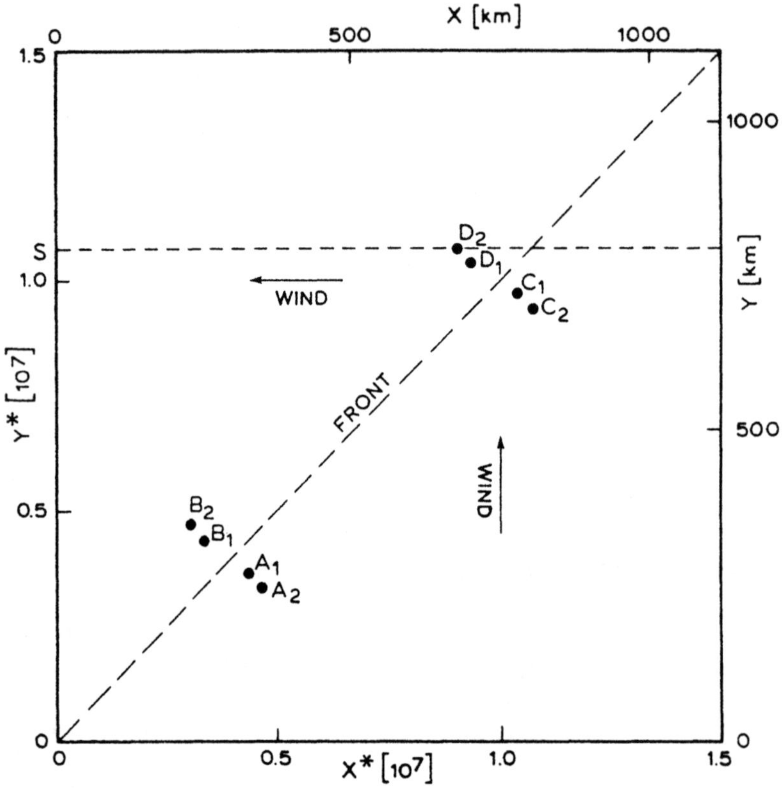

Fig. 11.1. Wind field geometry for Case V, diagonal front.

diagonal front increases from west to east, a progressively more energetic wave field inpinges on the front as one moves from SW to the NE along the front. These northward-traveling waves are transformed into swell after they have propagated across the front, where they can then interact with the new windsea developing to the west of the front. The directional spectrum at any location in the NW sector is determined by the superposition of the residual swell from the south and the new windsea which has grown from the front lying to the east. The processes affecting this balance are windsea growth under slanting fetch conditions, swell dissipation, and swell–windsea transfer by weak nonlinear interactions.

The resultant wave fields are inhomogeneous with respect to both spatial coordinates and strongly directionally dependent. Although this case is therefore considerably more complex than either Case III or Case VII, it is nevertheless of considerable interest, as it exemplifies the form of strong wind field discontinuities which may often be encountered in real frontal situations.

11.3. RESULTS

Figures 11.2 and 11.3 show two examples of the overall structure of the wave fields as summarized by the Custer diagrams for the distributions of the total wave energy and mean wave propagation direction. The models shown, BMO and DNS, belong to the same class, CD, and both incorporate some form of windsea–swell interaction (cf. Table 10.1). South of the front, the patterns for both models are similar: an essentially uniform growth downwind from the southern boundary up to the front, with some boundary effects visible at the eastern edge, and a still weaker influence of the oblique

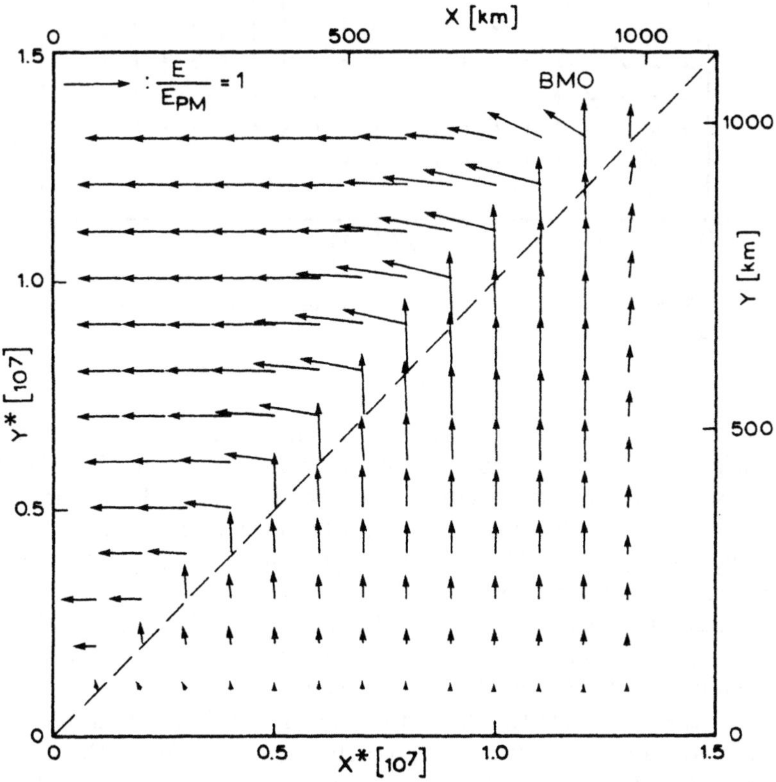

Fig. 11.2. Custer diagram showing E/E_{PM} and $\bar{\theta}$ for Case V for the CD model BMO.

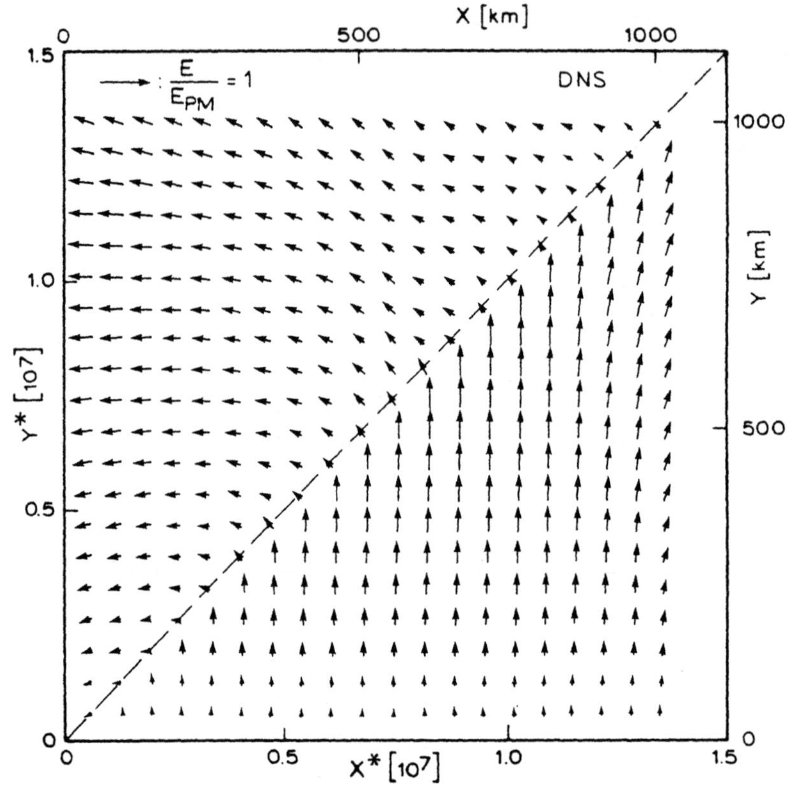

Fig. 11.3. Same as Fig. 11.2, except for the CD model DNS (note the different vector length scales) (Case V).

front near the front itself. Along the front and in the NW sector, however, the behavior of the two models is seen to be quite different. The relaxation of the windsea direction toward the local wind direction downwind (to the west) of the front is more rapid for the BMO model than for the DNS model, and the total energy fields also differ significantly. In the BMO model the wave energy increases across the front, while in the DNS model a large percentage of the energy present to the south of the front is lost when the waves cross the front. A similar energy deficit is apparent in the energy contour plot for the CH model NOWAMO (Fig. 11.4).

In general, the models show strong differences in the changes of total wave energy and other wave field properties across the front. These may be attributed to differences in the windsea–swell transfer algorithms, different slanting-fetch windsea growth rates for the newly generated windsea in the

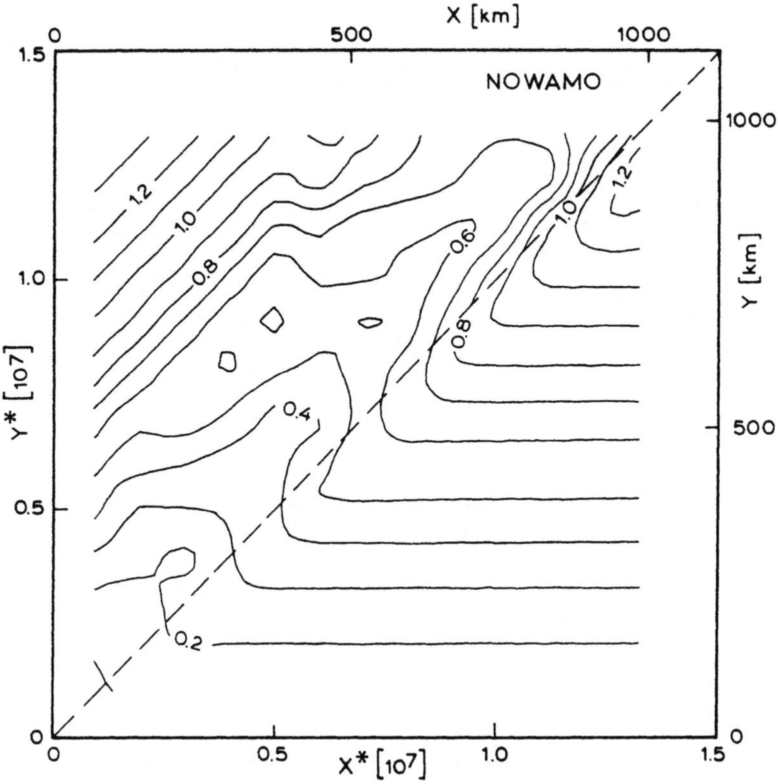

Fig. 11.4. Examples of energy contour field for Case V for the CH model NOWAMO.

NW sector, different swell damping rates, and numerical finite difference discrepancies associated with the discontinuities at the front. The model scatter is illustrated by Figs. 11.5 and 11.6, which show plots for each wave model of the total energy and wave direction, respectively, as functions of x^* along the EW section S at $y^* = 1.07 \times 10^7$ (cf. Fig. 11.1). Note that x^* decreases from left to right, so that westward wave growth is represented in the usual way with fetch increasing toward the right.

Three models (BMO, GONO, and HYPA) show a rapid increase in wave energy as the waves cross the front, three models (DNS, TOHOKU, and NOWAMO) show a strong decrease, while for the remaining three models (VENICE, MRI, and SAIL) the transition is relatively smooth (Fig. 11.5).

An increase in wave energy occurs in models which assume that (1) a significant portion of the northward-traveling wave energy is immediately transferred into windsea energy when the waves cross the front, with the

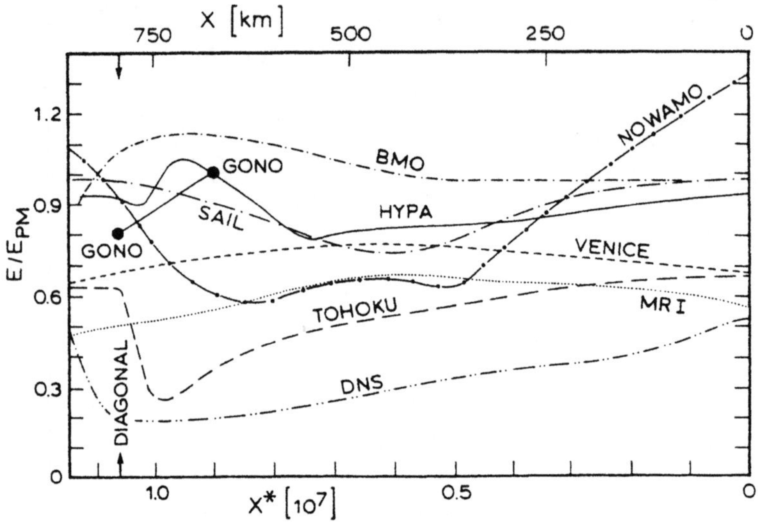

Fig. 11.5. Total wave energy along the section S (Case V, cf. Fig. 11.1). Note that fetch increases to the right (decreasing x^*).

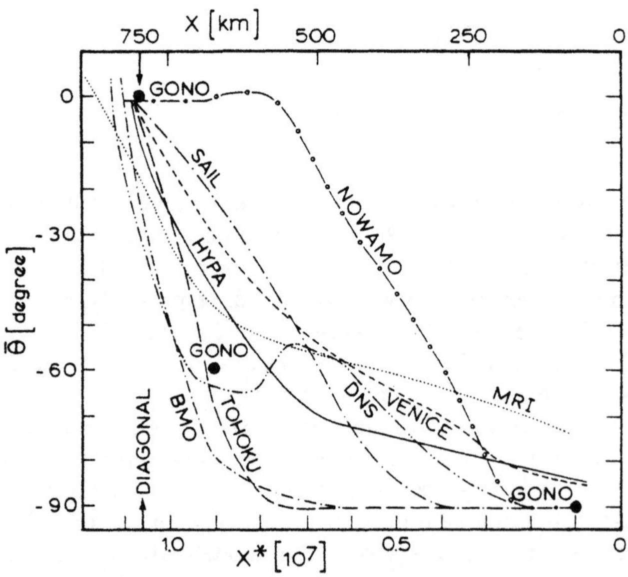

Fig. 11.6. Relaxation of mean wave direction for Case V along section S (cf. Fig. 11.1). Note that fetch increases to the right (decreasing x^*).

new windsea traveling either in the new wind direction (BMO, GONO) or in a northwesterly direction between the new and the old wind directions (HYPA), and (2) the remaining energy is retained without significant loss as swell. The immediately available windsea energy then provides a favorable initial state for subsequent rapid growth of the windsea, which is superimposed on an essentially constant residual swell. These models are also characterized by a rather rapid turning of the mean wave direction toward the new wind direction in Fig. 11.6.

The sudden decrease in wave energy across the front in the models DNS, NOWAMO, and TOHOKU has different causes.

In the DNS model, the swell–windsea transition algorithm limits all swell components with frequencies higher than the local wind frequency to a Pierson–Moskowitz saturation spectrum with a $\cos^2\theta$ spreading factor. This automatically removes all wave energy traveling in directions east of north, and strongly reduces the swell energy in the remaining westerly directions. The initial change in mean wave direction toward the west is correspondingly rapid. [The intermediate shift toward the north near $x^* = 0.7 \times 10^7$ is due to an eastern boundary effect; the residual swell which survives the frontal transition is particularly small in the NE region, where the windsea in the SE sector is veered toward the east (cf. Fig. 11.3). Thus in the central region of the section a relatively high proportion of residual NW swell is available to weight the mean wave direction further toward the north.]

The decrease of wave energy in the NOWAMO model may be attributed simply to the swell selection algorithm, which retains only a single directional component representing the highest energy density at each frequency.

In the TOHOKU model the energy loss arises from technical limitations of the numerical windsea advection scheme for discontinuous changes in the wind direction. These happen to be particularly marked for a front lying at 45° to the grid axes, as in the present case.

The DP models VENICE and MRI, finally, show a smooth transition across the front, as is to be expected for models with dynamically decoupled windsea and swell components, while the windsea–swell coupling algorithm of the CD model SAIL is sufficiently gentle that this model also shows a continuous transition. The slower directional relaxation of MRI compared with VENICE is due to the broader directional distribution and weaker swell damping rate of MRI.

For all models, the change in energy after the initial adjustments at the front is relatively gradual, the increase in new windsea energy being approximately balanced by the decrease in swell energy crossing the front as one moves westward along the section S.

A more detailed insight into the different mechanisms controlling the evolution of the spectrum after the waves have traversed the front is afforded by the 2d spectral contour plots shown in Figs. 11.7–11.9. The

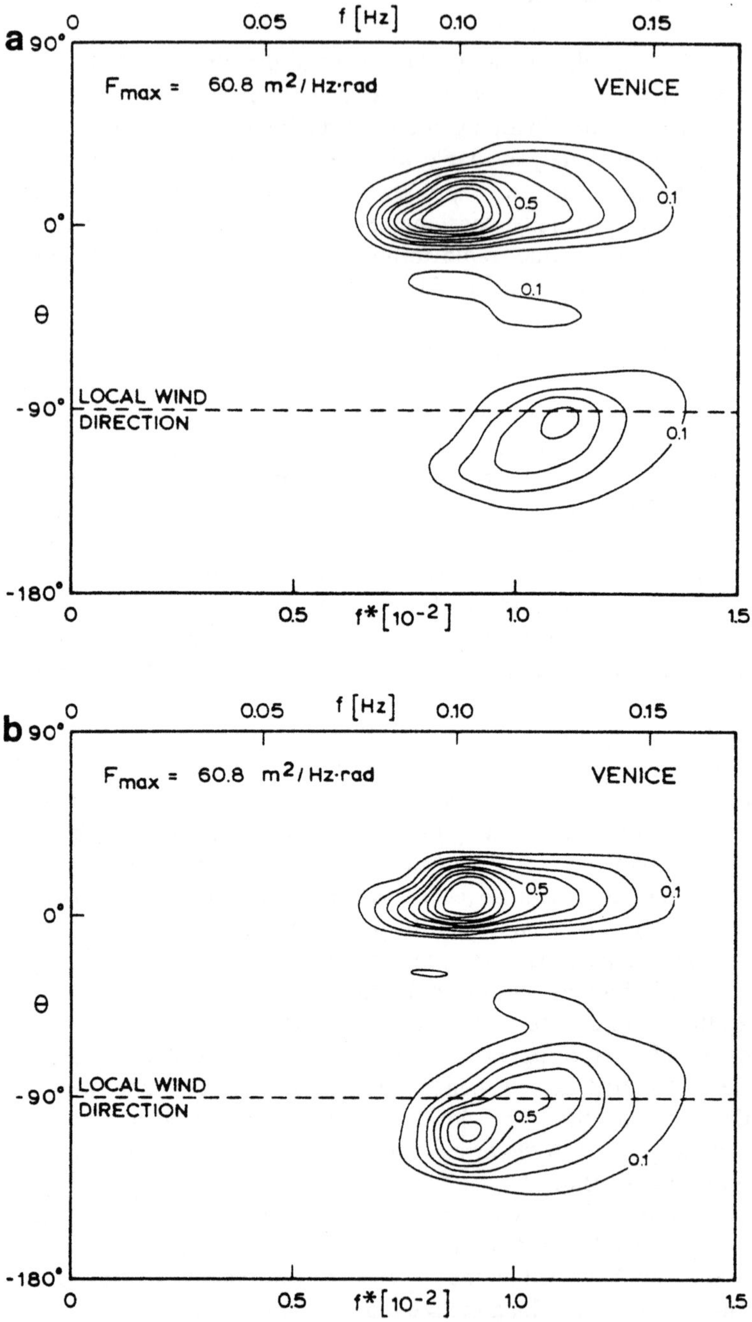

Fig. 11.7. Two-dimensional spectra for the DP model VENICE at points B_1 (panel a) and B_2 (panel b) close to the NW side of the front (Case V, cf. Fig. 11.1). Spectra are normalized by their peak values and contours are shown at intervals of 0.1.

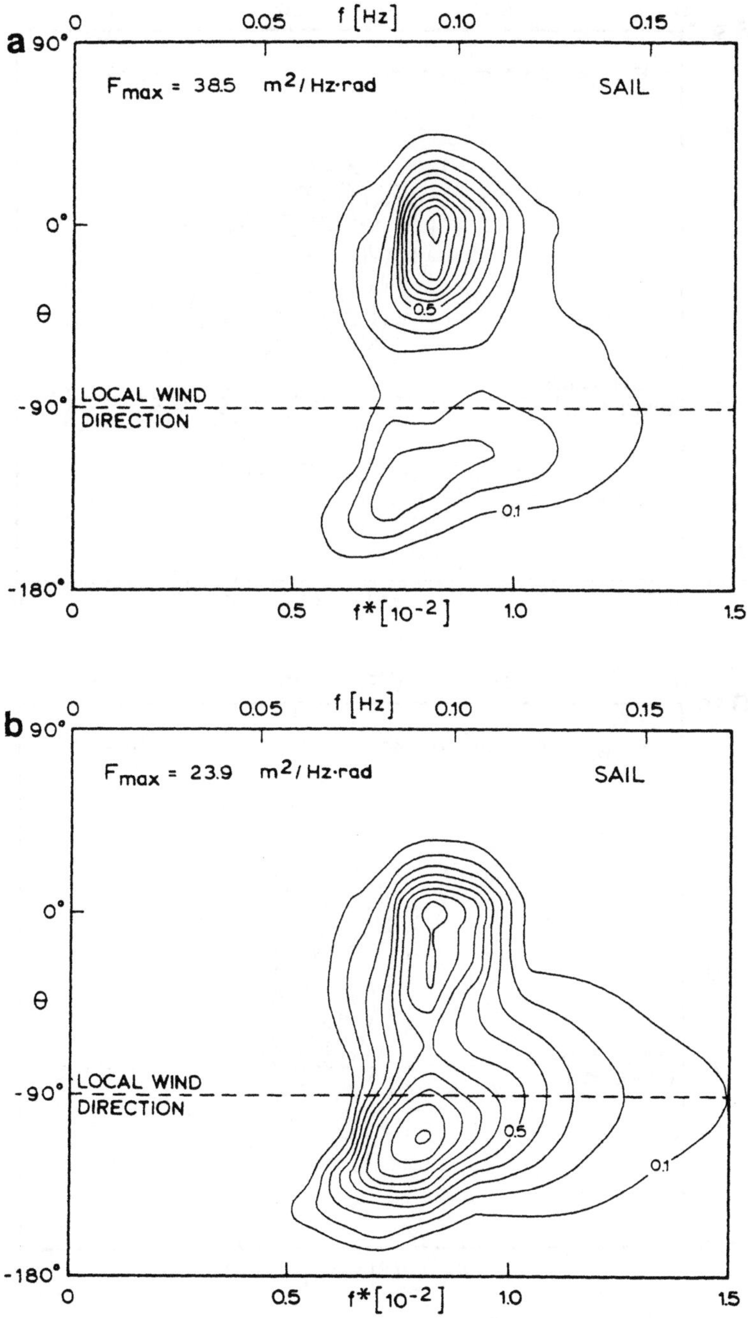

Fig. 11.8. Same as Fig. 11.7 except for the CD model SAIL (Case V).

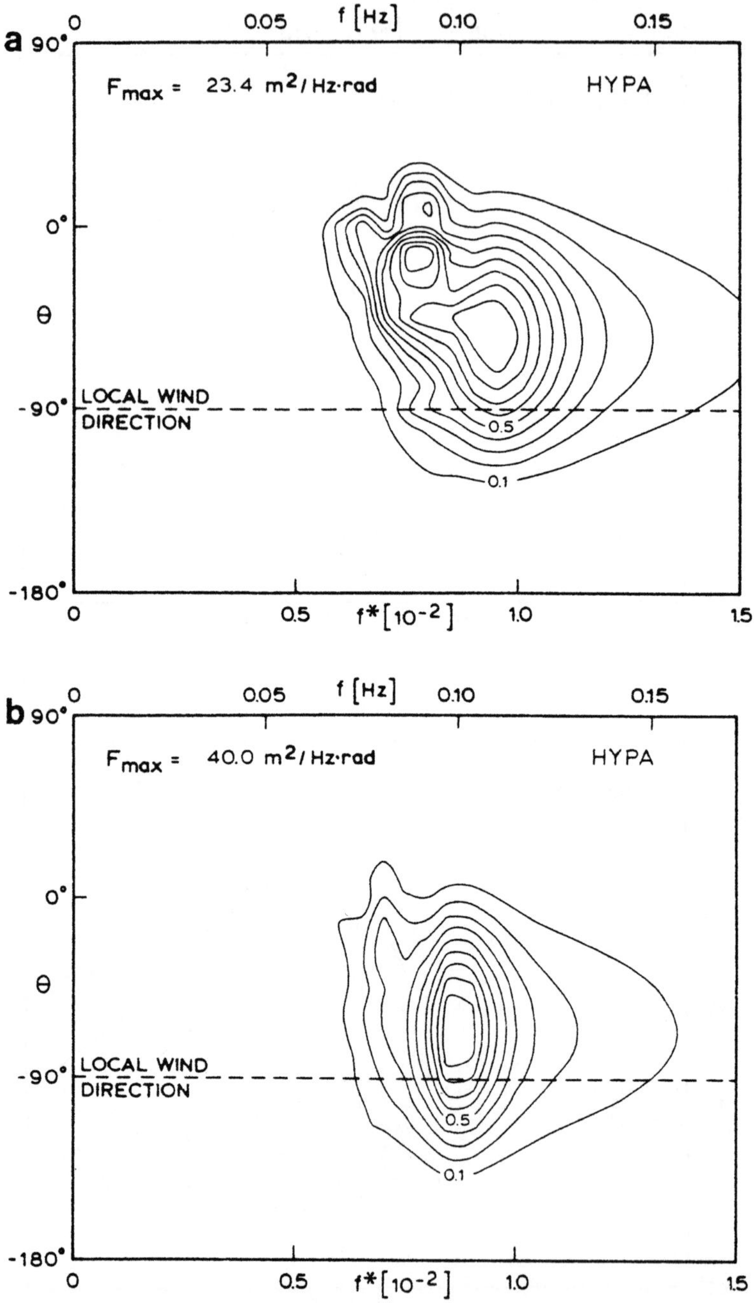

Fig. 11.9. Same as Fig. 11.7 except for the CH model HYPA (Case V).

figures show two successive spectra at the points B_1 (panels a) and B_2 (panels b) close to the NW side of the front (cf. Fig. 11.1) for three models selected from each of the three model classes: VENICE (DP), SAIL (CD), and HYPA (CH).

Typical for DP models is the superposition of clearly separated wind and swell regions in the VENICE spectra (Figs. 11.7a, b) with a rapidly growing, skewed windsea spectrum (cf. slanting-fetch Case III) and a slowly decaying swell spectrum. The CD model SAIL (Figs. 11.8a, b) exhibits qualitatively similar features, but the windsea–swell coupling algorithm leads in this case to a more rapid erosion of the swell under the given conditions of comparable windsea and swell frequencies. The CH model HYPA, finally is characterized by an already relatively well developed windsea at the point B_1 close to the front (Fig. 11.9a), and a subsequent rapid nonlinear absorption of the swell by the windsea (Fig. 11.9b). The swell–windsea transition algorithm of this model yields an initial mean windsea direction at point B_1 directed to the north of the local (westward) wind direction, with a subsequent relaxation to the wind direction.

11.4. CONCLUSIONS AND RECOMMENDATIONS

As in the duration-limited turning-wind Case VII (Chapter 10), the principal features of this case are governed by the way in which the models treat the windsea–swell transition regime. However, the regeneration of a new windsea beyond the front under slanting-fetch-limited conditions, as in Case III, and the propagation of a spatially inhomogeneous wave field into the front, create additional complexities which result in larger net differences between the models than in Cases III or VII (cf. Figs. 11.5 and 11.6). The results can be grouped broadly into the model classes DP (first-generation models with clearly separated, dynamically independent windsea and swell regimes) and CH and CD (second-generation models with strong windsea–swell interactions during the passage of the waves through the front, and further windsea–swell coupling in the interior regions of the NW sector when the windsea and swell frequencies become comparable, e.g., at point B_2). However, the details of the windsea–swell coupling do not readily group into the model classes CH and CD, but must be grouped with respect to the individual coupling criteria listed in Table 10.1.

The principal conclusion which may be drawn from this case study is that the behavior of the wave field propagating across a front is governed by processes which can in principal be computed within the framework of our present understanding of the spectral energy balance (for example by a model of the form EXACT-NL), but which are parameterized rather arbitrarily and in very different ways in present wave models. Improved parameterizations of the windsea–swell coupling clearly need to be developed and tested explicitly against exact computations of the nonlinear transfer under these special conditions.

Frontal passages are commonly occurring meteorological phenomena. In view of the widely varying and complex 2d spectral distributions predicted by different models under these situations, it would be useful to attempt to document the wind and wave fields for such events more completely than has normally been achieved in the past. Together with the (still simpler) slanting fetch Case III, Case V, if generalized to a moving front, represents a basic wind field which may often be encountered in nature in a reasonable approximation to the idealized geometry considered here.

Chapter 12

Case VI: Stationary and Moving Hurricane

12.1. WIND FIELD, BOUNDARY CONDITIONS

In this case two wind fields typical of intense hurricanes were investigated, the first representing a stationary storm, and the second representing the identical storm translating to the north at 15 m/s. The spatial distribution of wind speed and direction, typical for a strong northward-moving hurricane, is shown in Fig. 12.1. Also shown are the output grid points selected for display of the 2d spectra. The maximum wind speed of 40 m/s is located 50 km northeast of the eye. There is a pronounced asymmetry in wind speed (wind stronger to the east and north of the center) and a north–south asymmetry in wind direction (strong inflow in the rear quadrants).

The storm systems were defined on the domain $0 \leq x \leq 1300$ km, $0 \leq y \leq 1700$ km. The eye of the stationary storm is at coordinates (650, 1400) km (cf. Fig. 12.1). The eye of the moving storm is at the same coordinates after 24 h, and is therefore located 1296 km south of (650, 1400) at the start of the moving storm run.

The parametric model of Ross (1976) was used to provide initial and boundary wave conditions for both stationary and moving storm cases. This model defines the total energy and peak frequency at any point in a hurricane from the wind speed and radial distance from the eye. The directional spectrum is taken to be a JONSWAP-type spectrum with a $\cos^2\theta$ directional spread, assumed symmetrical about the local wind direction.

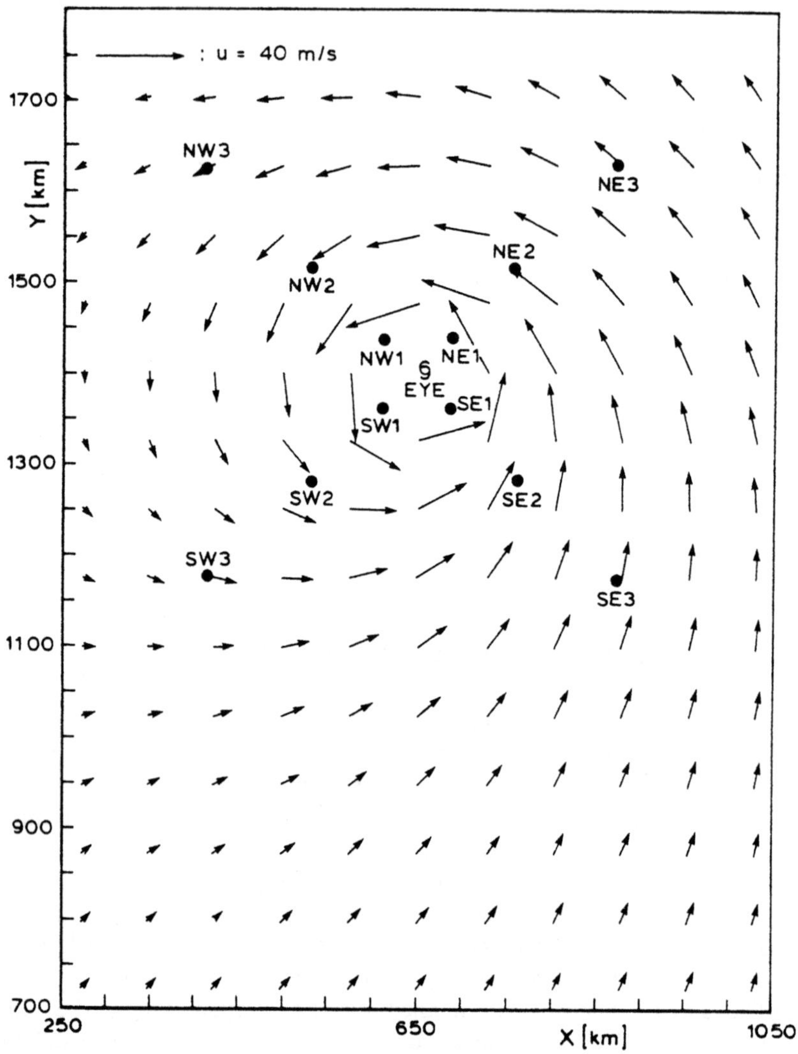

Fig. 12.1. Hurricane wind field for Case VI with selected output points for 2d spectra.

12.2. PURPOSE OF THE TEST

This test was designed to investigate the performance of models under rather extreme but nonetheless realistic conditions in which many of the processes considered independently in the previous case studies would come into play simultaneously.

Table 12.1
Wave Model Attributes for Case VI (Hurricane Wind Field)

	MRI	VENICE	NOWAMO	GONO	TOHOKU	HYPA	BMO	SAIL	DNS
Grid size									
x (km)	1280	3200	1300	1400	1320	780	1600	1350	800
y (km)	1720	4500	1700	1800	1720	1200	1680	2700	1000
Grid spacing (km)	40	50	75	40	40	30	40	37.5	40
Time step (h)	1	0.67	1.5	0.5	1	0.25	0.5	1	1
Boundary conditions	Yes	No	Yes	No	Yes	Yes	No	No	Yes
Initial conditions, moving storm	Yes (1/2 energy)	No	Yes	(Did not run)	Yes	No	No	Yes	No
Initial conditions, stationary storm	Yes (1/2 energy)	No	Yes	Yes	Yes	No	No	Yes	No

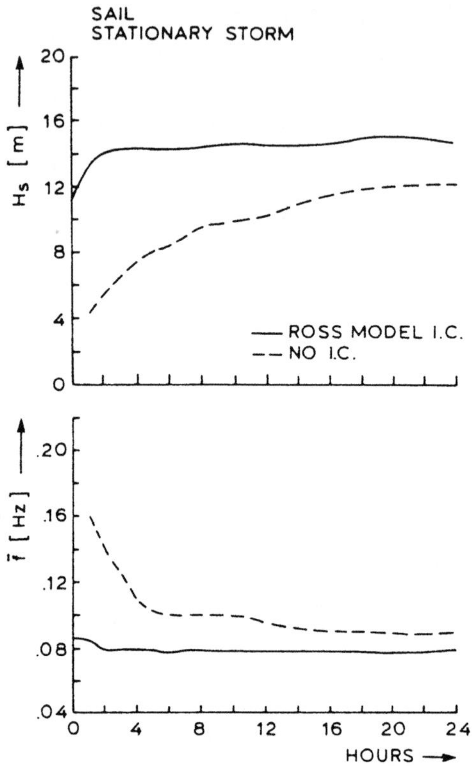

Fig. 12.2. Sensitivity of SAIL model hindcast of maximum H_s and associated $\bar{f}$ to initial conditions in Case VI.

The computational modifications and expense necessary to run this case precluded a uniform simulation by all modelers. Thus there was considerable variation among modelers in the choice of grid domain, grid spacing, time step, and treatment of boundary and initial conditions, as seen in Table 12.1. This (unavoidably) complicated the intercomparison of models. For example, there are indications from a sensitivity experiment conducted with the SAIL model (Fig. 12.2) that, without the initial state specification prescribed, steady-state solutions are not attained in the 24-h simulation conducted in this study, especially for the moving storm case. Thus we will emphasize here intercomparison of the stationary case and caution the reader that differences among models exhibited even in that case are at least in part due to differences in experimental conditions.

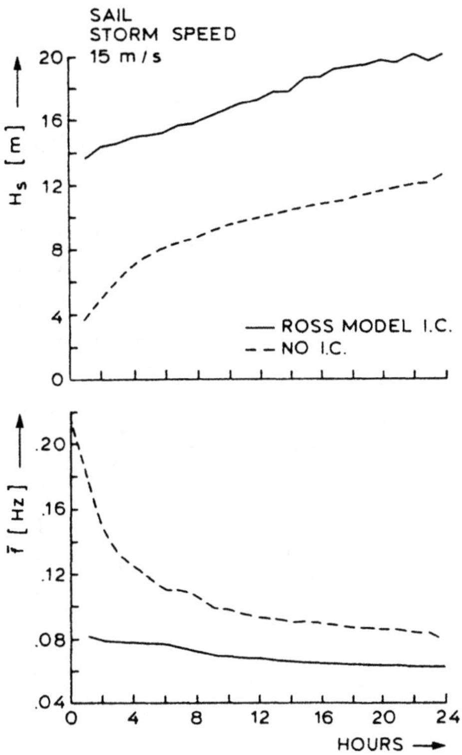

Fig. 12.2. (*continued*)

12.3. SPATIAL PROPERTIES OF WAVE FIELDS

The overall distributions of significant wave height (H_s) and mean direction ($\bar{\theta}$) for the stationary run are shown for four models in Fig. 12.3. The TOHOKU and NOWAMO patterns differ the most. In TOHOKU, high-sea states surround the eye and mean wave directions follow the local wind direction closely. The TOHOKU model requires the windsea to be in the wind direction (cf. Case III), hence its behavior was to be expected in the near field of the hurricane. A fast directional response and alignment of the windsea with the wind direction are also features of the discrete BMO model, which produced a similarly nearly circular Custer pattern in the inner region (not shown in Fig. 12.3). The NOWAMO model, on the other hand, exhibited a pattern of $\bar{\theta}$ least like the wind streamline pattern and showed high-sea states exclusively confined to the eastern side of the storm. This effect is

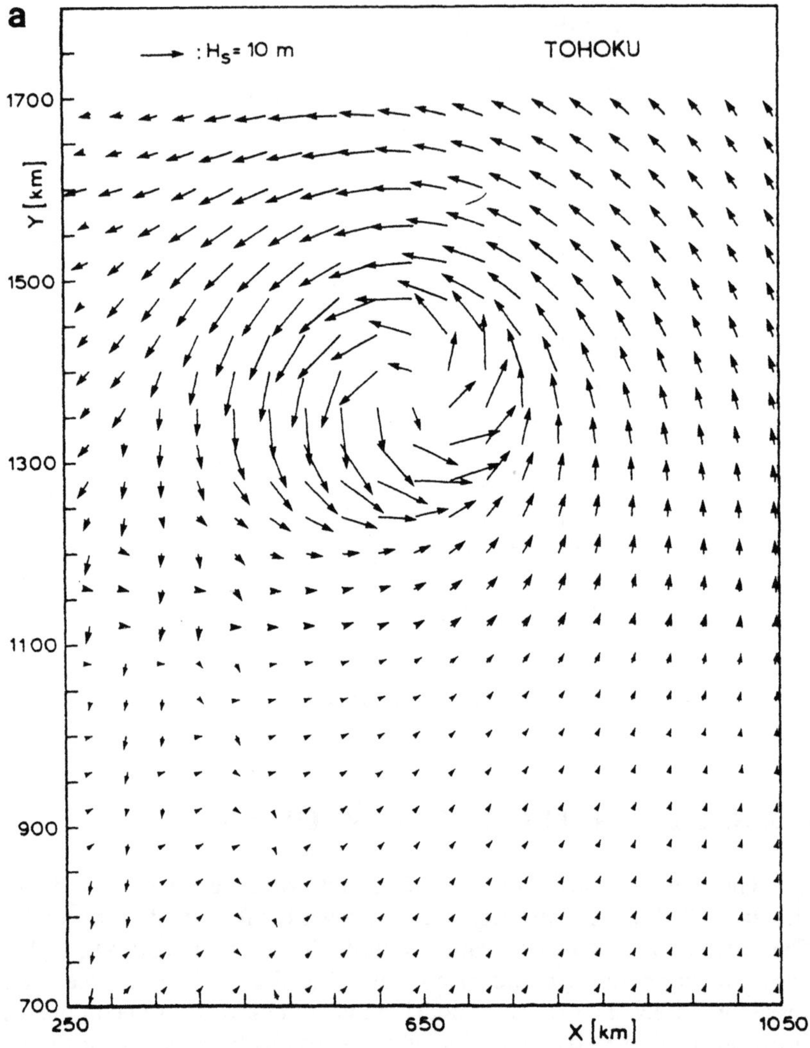

Fig. 12.3. Custer diagrams of significant wave height and direction for Case VI for the two CH models TOHOKU and NOWAMO (panels a and b), the CD model SAIL (panel c), and the DP model VENICE (panel d).

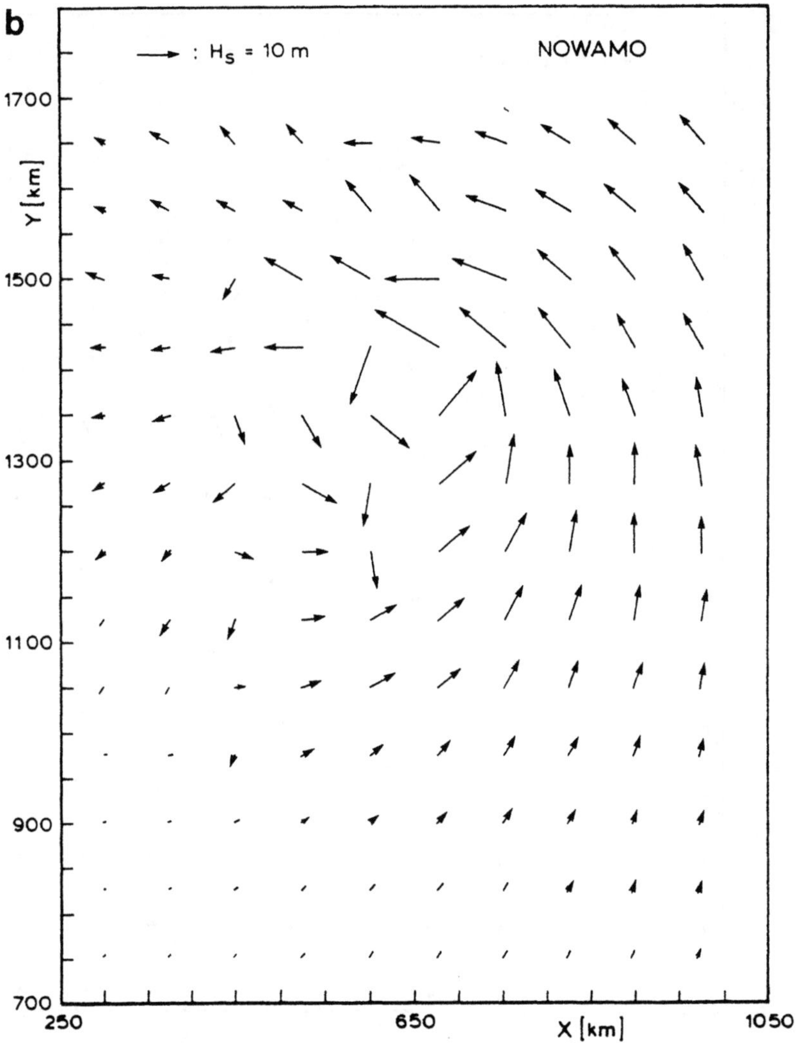

Fig. 12.3. (*continued*)

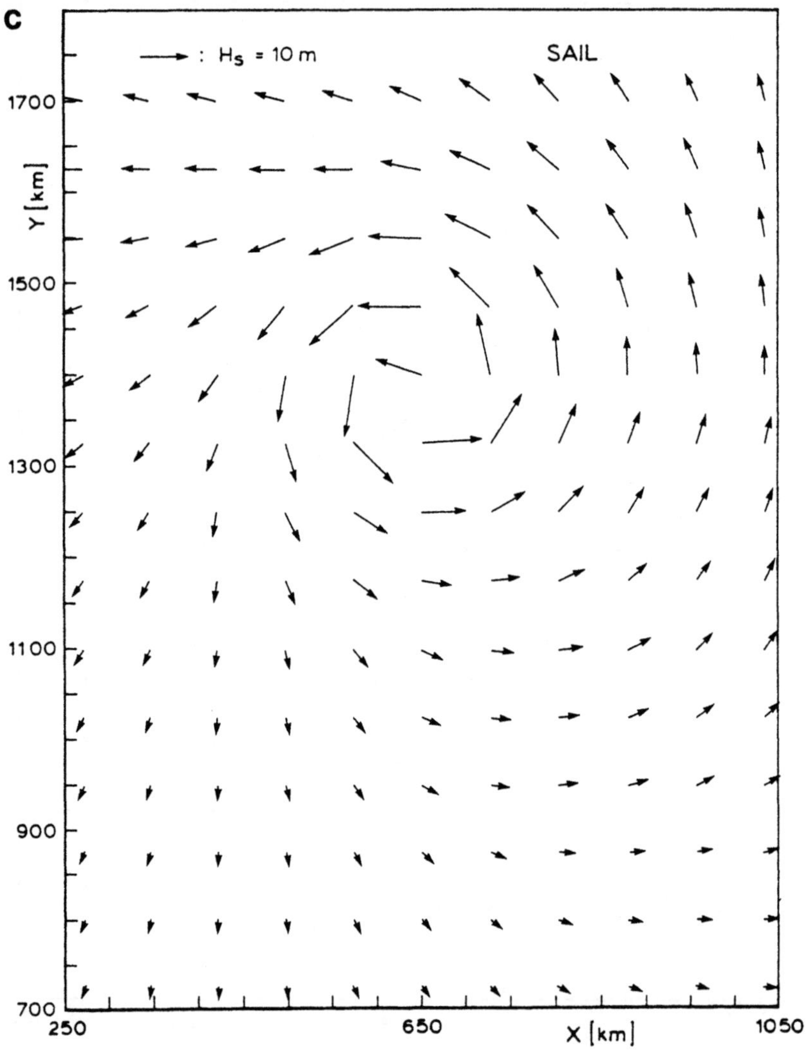

Fig. 12.3. (*continued*)

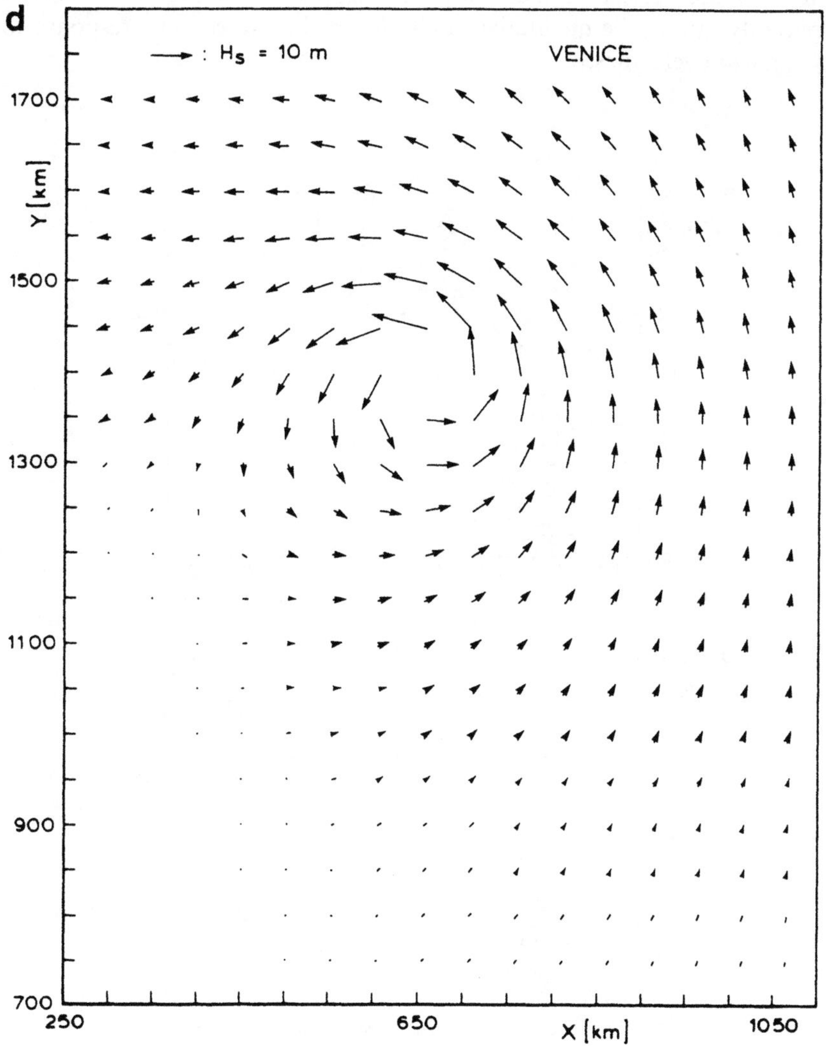

Fig. 12.3. (*continued*)

largely due to the nonstandard evaluation of swell in the NOWAMO model, which retains only the direction band containing the highest energy in a given frequency band. The other models produced patterns more or less similar to the VENICE or SAIL patterns, also shown in Fig. 12.3. This is apparently due to the qualitative similarity in their directional response to turning wind velocities.

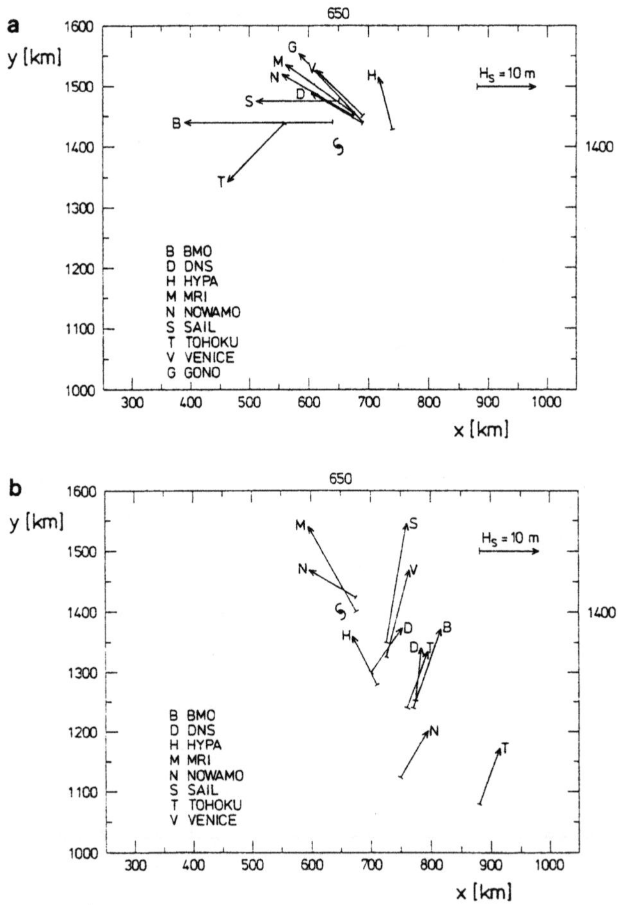

Fig. 12.4. Positions of $(H_s)_{max}$ for different models for the stationary hurricane (a) and moving hurricane (b) (Case VI). Arrows point in the mean wave direction $\bar{\theta}$ and are proportional to $(H_s)_{max}$ in length. (Results are not available for GONO for the moving hurricane).

Table 12.2
Wave Model Intercomparison Case VI (Hurricane Wind Field)

Stationary

	MRI	VENICE	NOWAMO	GONO	TOHOKU	HYPA	BMO	SAIL	DNS
H_s maximum (m)	14	11	14	16	13	8	25	14	10
Bearing (from eye)	NE	NE	NE	NE	W	NE	NE	NE	NE
Distance (km)	80	90	50	60	80	100	50	50	60
Distance $H_s/2$	325	240	450	250	260	> 500	150	400	> 500
$\bar{f}$ at $(H_s)_{max}$	0.08	0.09	0.11	0.08	0.08	0.10	0.07	0.08	0.10
$\bar{\theta}$ at $(H_s)_{max}$	NW	NW	NW	NW	S	NNW	NW	NW	NW

Moving

	MRI	VENICE	NOWAMO		TOHOKU		HYPA		BMO	SAIL	DNS
H_s maximum (m)	16	15	9		10		10		14	20	9
Bearing (from eye)	ESE	SE	NE	SSE	SE	SE	SE	SSE	SE	SE	SE
Distance (km)	60	100	50	275	175	400	120	220	200	90	200
Distance $H_s/2$	325	525	400	700	—	700	—	650	650	375	600
$\bar{f}$ at $(H_s)_{max}$	0.07	0.08	0.12	0.11	0.10	0.08	0.10	0.10	0.08	0.07	0.10
$\bar{\theta}$ at $(H_s)_{max}$	N	N	NW	NE	NNE	NNE	NNE	NNE	NNE	NNE	NNE

A parameter of considerable interest in hurricanes is the location and size of the maximum significant wave height $(H_s)_{max}$, with attendant wave parameters, e.g., $\bar{f}$ and $\bar{\theta}$. From contour plots of H_s, $\bar{f}$, and $\bar{\theta}$ we have extracted these parameters for both moving and stationary storms in Fig. 12.4 and Table 12.2. The following features are evident:

1. The values of $(H_s)_{max}$ range between 8 and 16 m for the stationary storm, except for BMO, which gives 25 m (Fig. 12.4). The moving storm model results also produce $(H_s)_{max}$ values in the range 9–16 m, except for SAIL, which yields 20 m. There is no consistent change in $(H_s)_{max}$ between the stationary and moving storms: three models exhibit higher wave heights for the moving storm, four models yield lower heights, and one model shows no change (Table 12.2; results for GONO for the moving storm are not available). Some of the differences may be due to the variety of initial and boundary conditions used in this test. The high BMO value for the stationary storm is due to the instantaneous adjustment of the windsea direction to the local wind direction. This creates a "resonance" effect for waves propagating along a high-wind streamline around the center of the eye. The waves are able to grow to a fully developed state despite the curvature of their infinite "fetch." The windsea of the one-parameter hybrid models also follows the local wind direction, but the adjustment of the windsea direction to the local wind direction in these models results in some energy loss through radiation to swell. The high values of the SAIL model for the moving storm result from a combination of the relatively high growth rates at higher wind speeds associated with the u_*-formulation of the SAIL growth expression and the fact that the longer waves generated in this manner in the SE quadrant experience a longer effective fetch through a quasi-group velocity resonance with the moving storm.
2. The mean frequency at $(H_s)_{max}$ also shows rather different behavior among models for the stationary and moving cases: three lower values for the moving storm, three higher values, and two models with no change (Table 12.2). However, the frequency values are generally consistent with the $(H_s)_{max}$ values and can be explained by the same mechanisms.
3. For most models, the location of $(H_s)_{max}$ shifts from a point near the hurricane eye in the NE quadrant for the stationary storm case to a point further away from the eye in the SE quadrant for the moving storm (cf. Table 12.2). The TOHOKU model places $(H_s)_{max}$ in the western sector of the stationary storm but locates it, as do the

other models, in the southeastern quadrant of the moving storm. The shift of $(H_s)_{max}$ to the SE quadrant for the moving storm may be explained by the well known group velocity resonance effect, which was noted as particularly marked in the SAIL model.

12.4. CHARACTERISTICS OF THE DIRECTIONAL SPECTRA

Only the stationary case will be discussed in this section. The 2d spectral contour plots were compared both in the inner storm area (points labeled NE1, NW1, SW1, SE1 in Fig. 12.1) and in the outer region (NE3, NW3, SE3, SW3). Figure 12.5 shows examples of the directional spectra at NE1 for the models MRI (DP), HYPA and GONO (both CH), and BMO and DNS (both CD).

The most apparent differences in the distributions are in the positions of the peak. The BMO and MRI models are peaked near 0.05 Hz, the DNS and HYPA models have higher peak frequencies at 0.075 and 0.08 Hz, and the GONO model has an intermediate peak frequency at 0.065 Hz. The low peak

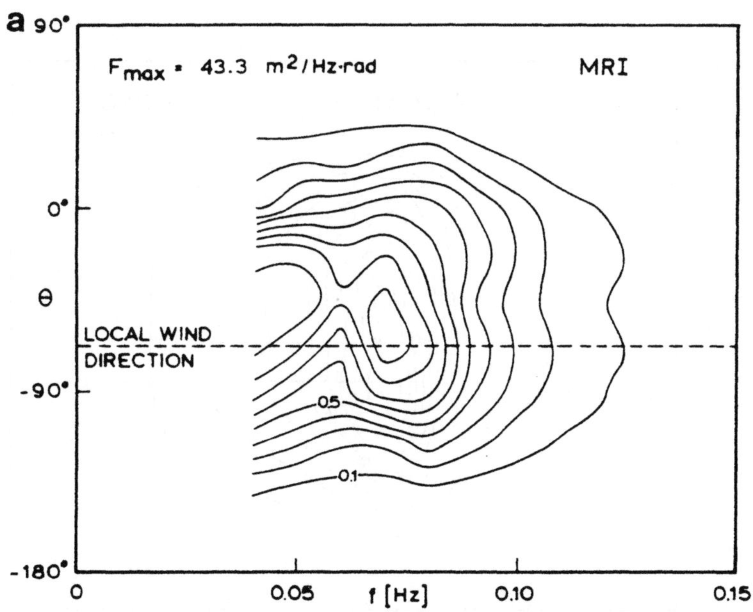

Fig. 12.5. Contour plots of 2d spectra at point NE1 of Case VI (cf. Fig. 12.1) for the DP model MRI (panel a), the CH models HYPA and GONO (panels b and c), and the CD models BMO and DNS (panels d and e). Contour intervals are 0.1 in units of F/F_{max}.

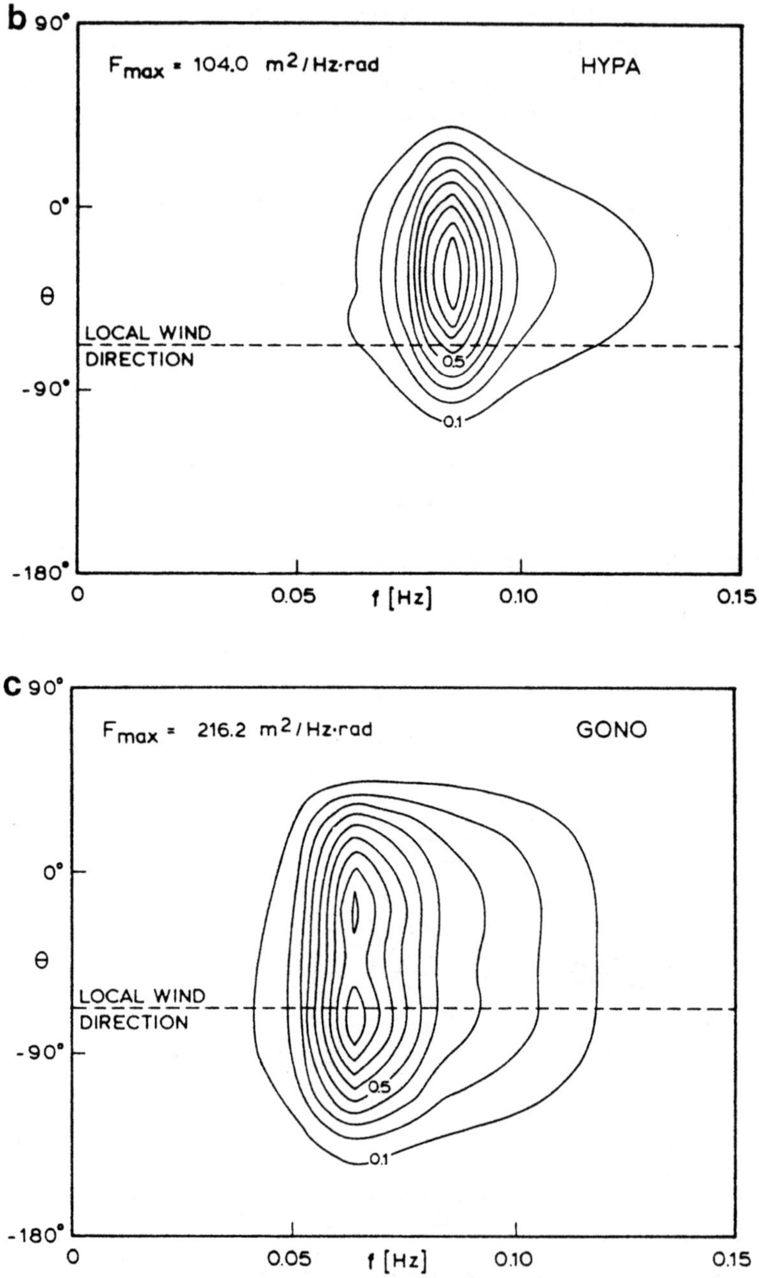

Fig. 12.5. *(continued)*

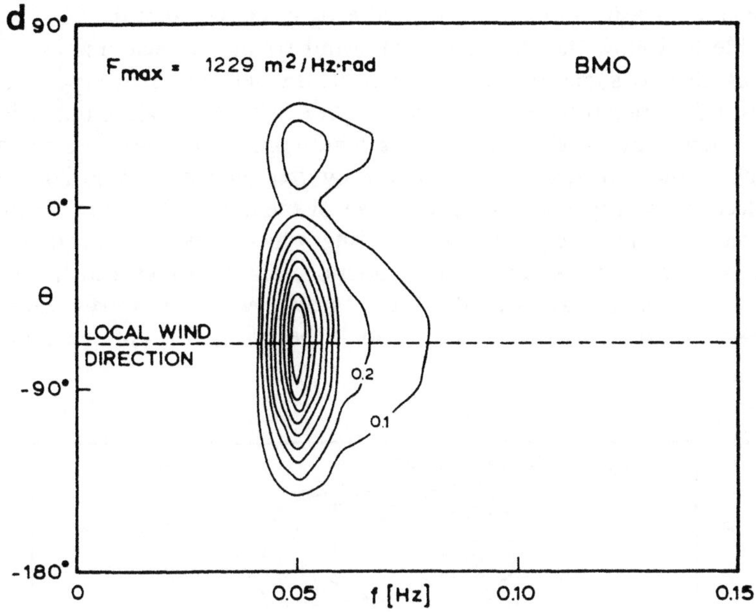

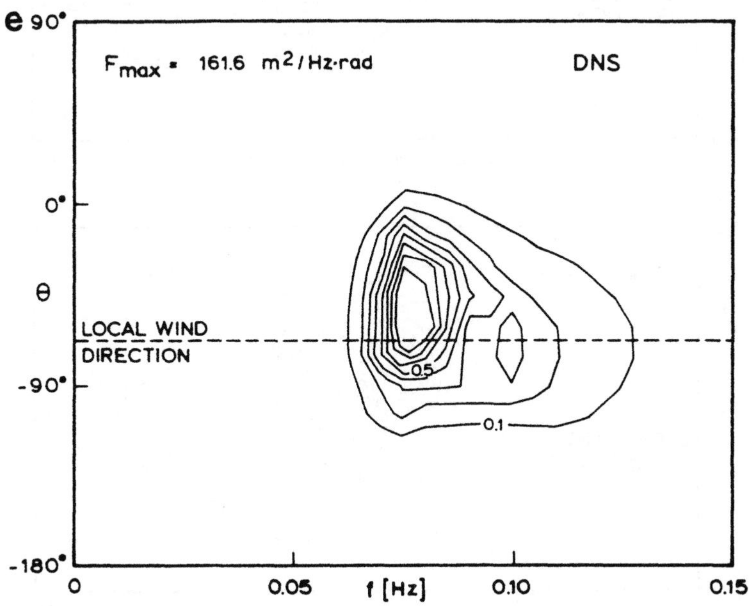

Fig. 12.5. (*continued*)

frequency for BMO is due to the fact, already discussed, that the wind waves are able to follow the curved circular wind streamline near the eye. This applies also to some extent for the GONO model, which incorporates no windsea directional lag effect, as in the HYPA and DNS models, but radiates more energy into swell under turning wind conditions than does the BMO model. In the MRI model, the relatively low frequencies are associated with the decoupled propagation characteristics of the model. The long waves are exposed to a relatively long fetch during which they are forced by the Phillips A term in the input source function. This provides a high starting level to trigger the exponential BF-type growth when the waves reach the high-wind region near the eye. (It should be noted that for the standard

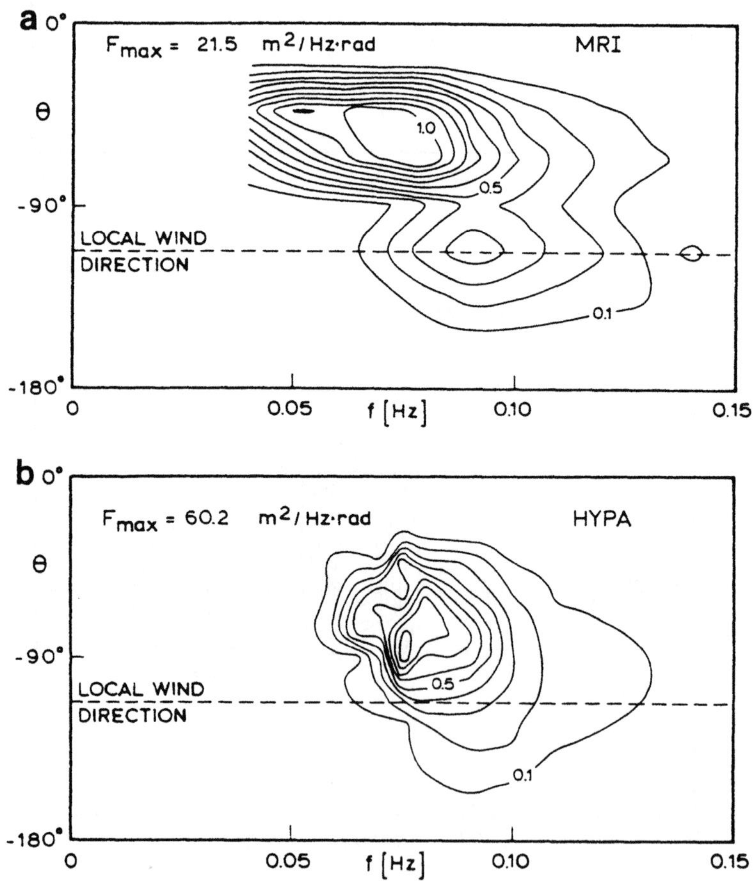

Fig. 12.6. Same as Fig. 12.5 at the point NW3 (Case VI, cf. Fig. 12.1).

fetch-limited uniform-wind case, the MRI model develops low frequencies more slowly than other models, cf. Case II.) The wave direction at the peak is identical to the wind direction in the BMO and GONO models, while the other models, which simulate directional relaxation effects, lag the wind by angles varying from 20° to 40°.

The greater flexibility of the discrete models DNS and MRI in representing the directional distribution relative to the hybrid models HYPA and GONO and the "quasi-hybrid" discrete model BMO is more apparent at locations, for example point NW3, for which the windsea and swell components are more clearly separated (cf. Fig. 12.6). The separation of the spectra into two (sometimes more) maxima representing the local windsea

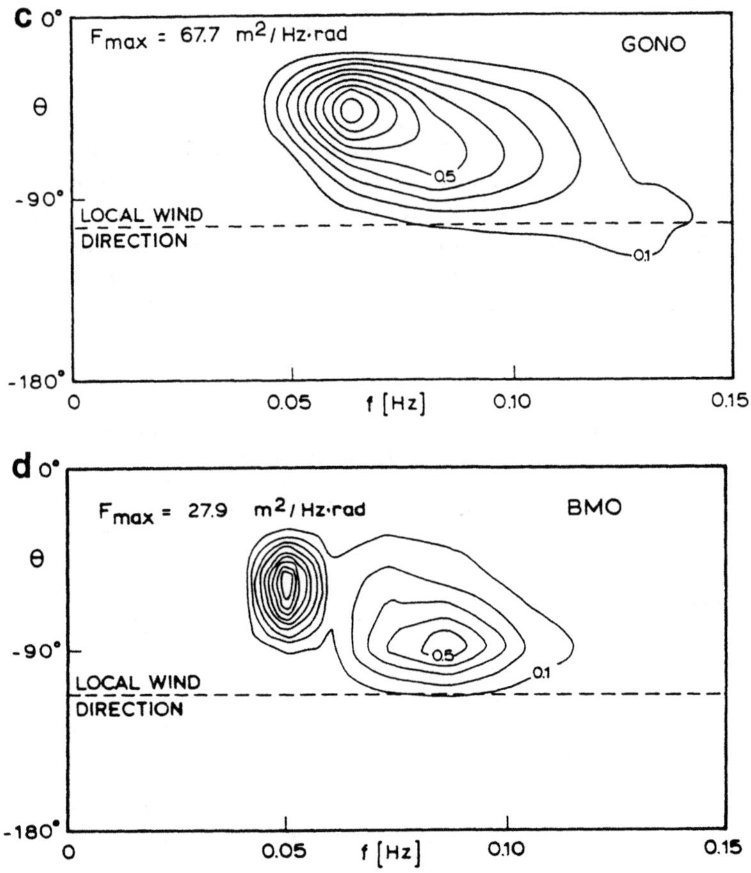

Fig. 12.6. (*continued*)

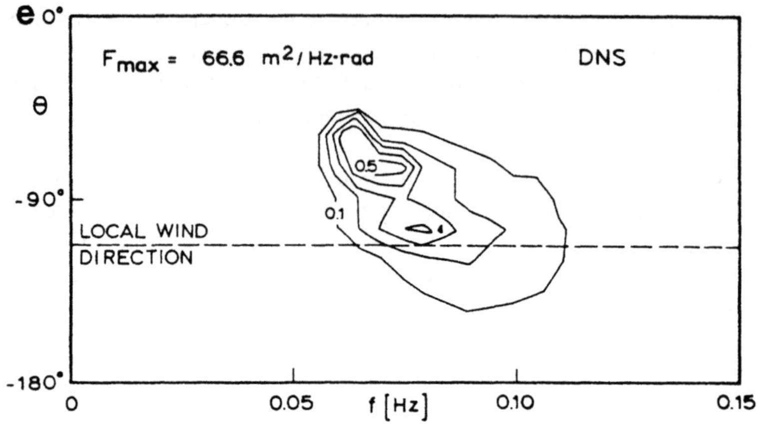

Fig. 12.6. (*continued*)

together with a swell contribution at a lower frequency and different propagation direction is typical of most locations. The swell generally propagates at angles of the order of 20°–40° off the local wind direction. Table 12.3 lists the propagation directions of the peak of the directional spectrum for all models at all inner and outer region grid points. The local wind direction at each point is also shown.

Table 12.3
Propagation Direction of Local Maximum of Directional Spectrum
Case VI Stationary Hurricane Wind Field

	Inner region				Outer Region			
	NE1	SE1	SW1	NW1	NE3	SE3	SW3	NW3
MRI	−45	30	I	−90	+25	110	−160	−45
VENICE	−30	30	170	−120	−30	15	−170	−90
NOWAMO								
GONO	−67	23	193	−105	+35	125	−145	−54
TOHOKU	−70	35	135	−145	−50	15	−165	−110
HYPA	−30	70	140	−110	−30	40	−160	−80
BMO	−70	20	110	−140	−50	115	−150	−60
SAIL	−30	30	130	−110	−30	120	−150	−90
DNS	−50	20	160	−80	−40	45	−160	−55
Local wind direction	−70	+20	+130	−140	−50	+15	+100	−110

[a] All angles in degrees positive clockwise from north.

The large differences in the frequencies and energy levels of the swell and windsea components in Fig. 12.6 and Table 12.2 may be attributed to several factors:

1. Differences in the properties of the windsea spectra in the inner region of the right front quadrant (see Fig. 12.5), which is the source area for this swell. The MRI spectra, for example, show considerable energy at low frequencies both at NE1 and NW3.
2. Differences in the windsea–swell transition algorithms. For instance, the TOHOKU model, when converting windsea to swell, assigns the new swell energy to a single direction band. In this case the appropriate direction band apparently "missed" NW3 (cf. Table 12.3).
3. The treatment of directional relaxation in a turning wind (cf. Cases VII and V).
4. The width of the directional distribution of the windsea, which affects the amount of energy radiated into swell when the wind turns (cf. Case IV).

12.5. CONCLUSIONS

Whereas the previous test cases were designed to isolate the influences of specific elements of the wave models, Case VI exercises all components of the wave models simultaneously. This makes straightforward "explanations" of model behavior difficult. However, the results of the case study may be summarized as follows:

1. Differences in the magnitudes and details of the spatial patterns of the H_s and $\bar{f}$ distributions were found to be large. The response patterns could not be readily grouped by model class. Qualitatively, all models reproduced the basic structure of the expected wave field in terms of, for example, regions of highest waves, general wave directions, and superposition of windseas and swell, irrespective of model class. But significant quantitative differences arise through differences in detail in the simulation of the individual processes.
2. The maximum energy level in the inner region of the stationary storm is very sensitive to the form of coupling between the wind and wave directions in the windsea growth algorithm. The highest wave

heights were excited in models which exhibit rapid directional relaxation and weak radiation into swell under turning wind conditions. For this reason the maximum wave parameters in the stationary storm did not correlate very strongly with the model growth rates in Case II.

3. In the outer regions of the stationary storm, nearly all models produced a complicated directional spectrum consisting of locally generated high-frequency wind waves and lower-frequency swell radiated from the high-wind inner region of the hurricane. The peak frequencies and energy levels of the radiated swell varied widely among models because of the large differences in the spectra predicted in the inner region and differences in the way in which sea and swell are separated.
4. An important feature captured by all models was the shift of maximum wave heights from the northeast to the southeast storm quadrants between the stationary and moving storm cases (moving storm results were not available for GONO). This effect is due to the well known group velocity quasi-resonance, which occurs when the storm propagation speed matches the wave group velocity.
5. In contrast to the other test cases, there was considerable variation among models in the choice of grid domain, spacing, and time step and in the treatment of initial and boundary conditions. These variations probably caused some of the differences seen in the stationary run, and we believe that they prevent a conclusive intercomparison of the moving storm case.

The case study revealed considerable sensitivity to many model features. A specific model intercomparison study for a real hurricane using recently available hurricane wind field and directional wave spectra data (cf. Ross and Cardone, 1978) would therefore be a valuable exercise.

Chapter 13

Summary

A series of wave models representing three model classes [decoupled propagation (DP), coupled hybrid (CH), and coupled discrete (CD)] have been intercompared for seven test cases designed to emphasize different features of the models. The study has provided a comprehensive view of the present state of wave modeling and has revealed a number of critical areas in which further work is necessary. The principal conclusions may be summarized as follows:

1. First-generation DP models and second-generation CH and CD models, which are based on fundamentally different pictures of the spectral energy balance, yield significantly different relations between space and time variables in the development of a windsea. It is therefore impossible to tune a DP model to agree with a CH or CD model for all types of wind fields. If the models are tuned to agree for fetch-limited growth, DP models will develop faster than CD or CH models in the duration-limited case by a factor of the order of 1.5.

2. DP models also differ from CH or CD models in the shape of the windsea spectrum. For the standard fetch- or duration-limited growth cases, DP models yield a nested family of spectral growth curves with no overshoot. For more complex wind fields, DP models generally develop a wider variety of spectral distributions than CH or CD models. Because different wave components evolve independently along different rays, the evolution of the spectrum in DP models is generally more sensitive to fetch variations with propagation direction. However, this is strongly dependent on the assumed directional distribution of the growth coefficients. In the cases

studied, the effect was clearly pronounced in the DP model MRI, which used relatively broad $\cos^2\theta$-dependent growth coefficients, but was insignificant for the second DP model VENICE, which assumed a narrower angular distribution for the growth coefficients.

3. Despite the basic incompatibility of first- and second-generation wave models, DP models remain attractive because of their greater computational simplicity. They can be useful if adapted and applied to a particular class of wave conditions. In the case of the two DP models considered in the present study, the MRI model is used primarily for open-ocean, duration-limited conditions, while the VENICE model was originally designed for the typically fetch-limited wave conditions of the Adriatic.

4. All present second-generation models suffer from limitations in the parameterization of the nonlinear energy transfer, S_{nl}. This process largely controls the shape and growth of the windsea spectrum. The models perform satisfactorily for the standard fetch- and duration-limited growth situations for which the parameterizations were designed. This applies also to slowly varying large-scale wind fields, for which the internal nonlinear shape adjustment time of the spectrum is small compared with the characteristic time scale of the wind changes experienced by the waves. However, the case studies were designed to test the models under rather extreme conditions of rapidly changing winds. For these situations, the parameterizations of S_{nl} generally contained too few degrees of freedom to cope with the wide variety of spectral distributions which may arise. In view of this shortcoming, which is common to both types of second-generation model, the potential advantage of CD models in providing a complete discretized 2d description of the windsea spectrum, as opposed to the parametrical description of CH models, could not be fully realized.

5. In the present stage of development, a CH model including a prognostic wave direction variable $\bar{\theta}$ (HYPA) generally yields predictions comparable to CD models with respect to directional relaxation effects. CD models have the advantage of predicting more detailed (but not necessarily correct) windsea directional distributions. As is to be expected, models which include no windsea directional relaxation show deficiencies for wind fields involving rapidly changing wind directions.

6. Many of these basic features distinguishing models of different classes become clearly recognizable in the case study analyses only if the individual model results were normalized in terms of the corresponding fetch-limited growth data of the model. Without this normalization, the variations in the basic fetch-limited growth rates of different models may

mask the variations associated with different wind field geometries. This result was rather surprising in view of the fact that the calibration of most models rests heavily on fetch-limited growth data, and more data exist for this growth case than for any other growth situation. We attribute the differences in large part to uncertainties as to whether the wind speed (U_{10} or $U_{19.5}$) or the friction velocity u_* is the more appropriate parameter to characterize wave growth, and to the scatter in published drag coefficients relating the two. A contributing factor may be the uncertainty in the boundary layer formulation used to relate these surface velocities to the geostrophic wind, which is normally provided as the primary input by the atmospheric models used to drive the wave model. Other factors, such as the gustiness of the wind, may also contribute to the discrepancies.

Formal agreement on some standard fetch-limited growth law (whether correct or not) would facilitate future intercomparison studies and aid model development by providing a generally accepted reference baseline.

7. Another significant source of divergence in the model results which is not related to model class is the choice of directional distributions. This is seen most clearly in test cases involving the transition from windsea to swell in a turning or inhomogeneous wind field. Existing measurements of the frequency dependence of the directional spreading factor for a fetch-limited growing windsea are not incorporated into any of the models except SAIL, presumably because it is felt that a simple frequency-independent spreading factor is adequate for most applications. The test cases indicate that a more sophisticated treatment of the directional distribution is called for in all cases in which windsea directional relaxation effects come into play. However, the correct simulation of fetch-limited spreading factors is only a first step in this problem. An adequate treatment requires a more general parameterization of the nonlinear transfer, which controls the energy redistribution in a turning windsea and in the swell–windsea transition regime. A more detailed analysis of existing slanting-fetch growth data would provide a valuable test of the directional response of models.

8. Finally, the superposition of these many differences, related either to model class or to individual model assumptions, can result in substantial net discrepancies among different model predictions for complicated wind fields, such as the hurricane wind fields of Case VI. However, it should be recalled that the test wind fields were chosen to reveal and emphasize critical elements of the models. For more slowly varying wind fields more typical of the average case, the models may be expected to show greater agreement. On the other hand, wave models are frequently needed just for extreme-event forecasts and statistics. Our intercomparison study suggests

that the reliable performance of the models in these situations, for which direct measurements are unfortunately sparse, must presently be regarded as highly questionable. A detailed analysis of the few existing hurricane wind and wave data sets in the light of the model intercomparison results for Case VI would clearly be valuable.

By identifying the most critical sources of these discrepancies, the model study has revealed several avenues which could be profitably explored to remedy the present situation and improve the state of wave modeling.

Chapter 14

Outlook

Ideally, a numerical wave model should compute the 2d wave spectrum, starting from a postulated functional form of the three basic source function constituents S_{in}, S_{nl}, and S_{ds}, for a prescribed wind field and boundary conditions, without any prior information on the form of the resultant wave spectrum. In the present study, only the EXACT-NL model integrated the transport equation in this manner (with respect to one integration coordinate only). The first-generation DP models compute the initial growth rate from prescribed source functions, but presume a given limiting form for the equilibrium spectrum. The second-generation CH models assume a given quasi-equilibrium shape (or family of shapes) for the entire windsea spectrum, predicting only one, or at most a few, characteristic windsea parameters. Finally, attempts to integrate the full transport equation in second-generation CD models using simple parameterizations of S_{nl} lead to spectral instabilities at frequencies beyond the peak frequency, so that these models also have to assume a prescribed spectral shape over much of the windsea spectrum.

It may be hoped that these difficulties will be removed in a third generation of CD models which employ discretized continuous-operator parameterizations of S_{nl} containing the same number of degrees of freedom as used in the discrete representation of the spectrum. First versions of such parameterizations have been successfully tested in simple transport equation integrations. Once such parameterizations are available, the remaining unknown source functions S_{ds} can be readily tuned to yield model spectral growth characteristics consistent with observations for the standard fetch-limited case. [S_{in} may be regarded as given to first order by the measurements of Snyder *et al.* (1981).] While it cannot yet be predicted whether such models will be able to resolve all difficulties encountered in the test cases of this report, a consistent differential approach to wave modeling, in which all hypotheses are expressed in terms of the source functions, rather than in the present mixture of source functions and the resultant wave spectra themselves, is clearly a prerequisite for removing the arbitrariness characterizing present parameterizations of the nonlinear interactions for complex seas.

It may be prophesized that CH models, even if extended to more parameters, will not be able to compete with future CD models when applied to extreme, nonuniform wind field conditions, such as those considered in the present case studies. Their principal value lies in the computational economy offered for the large majority of wind fields for which the windsea spectrum is able to adjust to a local equilibrium in a time period that is short compared with the appropriate wind field time scale. Our case studies suggest that a two-parameter model based on an energy (or frequency) scale parameter and a mean wave direction as prognostic variables should capture the essential features of the windsea development for such wind fields.

However, the windsea–swell transition regime may be expected to remain a problem area for CH models even for slowly varying wind fields. The basic difficulties associated with modeling the transition from strong to negligible coupling in CH models were partially masked in the present study by comparable shortcomings of present CD models in this region of the spectrum. While the problems of CD models in this respect are in principle resolvable, the difficulties of CH models are unavoidably associated with the basic separation of the spectrum into a decoupled swell spectrum and a strongly coupled windsea spectrum. To treat the windsea–swell transition region satisfactorily, a more sophisticated representation and parameterization of the nonlinear coupling in this region will be required.

Future CH models could conceivably be used in tandem with CD models, the CH models yielding first-iteration wave fields for most regions of the ocean, while the CD models would provide more reliable solutions in limited regions of strongly varying wind fields. Such applications may become relevant with the advent of future remote-sensing ocean satellites, when fast wave models will be required for the reconstruction of mutually consistent wind and wave fields from large, nonsynoptic data sets of global wind and wave measurements.

Appendix A

SWAMP Participants

Model	Institution	Participant (index to address)
MRI	Meteorological Research Institute, Japanese Meteorological Agency, Tsukuba, Japan	Takeshi Uji (1)
VENICE	Istituto per lo Studio della Dinamica delle Grandi Masse, Venice, Italy Associazione Studio Correnti Onde, Padua, Italy	Luigi Cavaleri (2) Luciana Bertotti (3)
SAIL	Atlantic Oceanographic and Meteorological Laboratory, National Oceanic and Atmospheric Administration, Miami, Florida, USA	Vincent Cardone (4) Linda Lawson (5) Arthur Greenwood (4) Robert Long (5)
DNS	Scripps Institution of Oceanography, La Jolla, California, USA Naval Ocean Research and Development Activity, NSTL Station, Mississippi, USA	Tim Barnett (6) James Allender (7) Mike Lybanon (7)
BMO	British Meteorological Office, Bracknell, United Kingdom	Brian Golding (8) James Ephraums (8)
GONO	Royal Netherlands Meteorological Institute (KNMI), De Bilt, The Netherlands	Gerbrand Komen (9) Willem de Voogt (10) Jan Bruinsma (11)
NOWAMO	Norwegian Meteorological Institute, Bergen, Norway	Johannes Guddal (12) Einar Maeland (12)

(*continued*)

Model	Institution	Participant (index to address)
HYPA	Institut für Meereskunde, Hamburg, Federal Republic of Germany	Heinz Günther (13) Wolfgang Rosenthal (13) Holger Linné (13)
TOHOKU	Geophysical Institute, Tohoku University, Sendai, Japan	Yoshiaki Toba (14) Sanshiro Kawai† Paimpillil Joseph (15)
EXACT-NL	Max Planck Institute of Meteorology, Hamburg, Federal Republic of Germany	Klaus Hasselmann (16) Susanne Hasselmann (16)

†Deceased 21 November 1981.

Addresses

1. Meteorological Research Institute, 1-1 Nagamine, Yatabe-machi, Tsukuba-gun, Ibarakiken 305, Japan
2. Istituto per lo Studio della Dinamica delle Grandi Masse, Consiglio Nazionale delle Ricerche, Palazzo Papadopoli, 1364 San Polo—30125 Venice, Italy
3. ASCO, Via Beato Pellegrino 14/20, 35100 Padua, Italy
4. Oceanweather Inc., 5 River Road, Suite 1, Cos Cob, Connecticut 06807, USA
5. AOML, NOAA, 4301 Rickenbacker Causeway, Miami, Florida 33149, USA.
6. Climate Research Group, Scripps Institution of Oceanography, University of California, La Jolla, California 92093, USA
7. NORDA, NSTL Station, Mississippi 39529, USA. *Present address for J. H. A.*: Chevron Oil Field Research Company, P. O. Box 446, La Habra, California 90631, USA
8. British Meteorological Office, London Road, Bracknell, Berkshire RG12 2SZ, United Kingdom
9. KNMI, P. O. Box 201, 3730 AE De Bilt, The Netherlands
10. Delft Hydraulics Laboratory, Laboratory "De Voorst," P. O. Box 152, 8300 ADS Emmeloord, The Netherlands
11. Ministry of Transport and Public Works–Rijkswaterstaat, Deltadienst, van Alkemadelaan 400, 2597 AT The Hague, The Netherlands
12. Norwegian Meteorological Institute, Vervarslinga på Vestlandet, Allégaten 70, 5000 Bergen, Norway
13. Institut für Meereskunde der Universität Hamburg and Max-Planck-Institut für Meteorologie, Bundesstrasse 55, 2000 Hamburg 13, Federal Republic of Germany
14. Geophysical Institute, Tohoku University, Physical Oceanography Laboratory, Aramaki-Aza-Aoba, Sendai 980, Japan
15. Center for Earth Science Studies, Regional Center, Kerala State, 682018 India
16. Max-Planck-Institut für Meteorologie, Bundesstrasse 55, 2000 Hamburg 13, Federal Republic of Germany

Appendix B

Consistency Requirements for Propagation Operators in Discrete Spectral Models

The numerical implementatation of the advection equation $(\partial/\partial t + \mathbf{v}\cdot\nabla)S = 0$ for a scalar field $S(\mathbf{x}, t)$ has been the subject of numerous investigations. The goal is normally to devise a numerical finite difference scheme for a cartesian or some other discrete grid which simulates the propagation of an initial field (for example, a single grid point "δ-function") at velocity $\mathbf{v}$ with minimal dispersion. However, for the more general radiative transport equation $(\partial/\partial t + \mathbf{v}\cdot\nabla)\ F(f, \theta; \mathbf{x}, t) = 0$, in which the group velocity $\mathbf{v}$ is a function of frequency f and propagation direction θ, it is not desirable to minimize the dispersion of the numerical advection operator if a discrete representation of the spectrum is used.

The representation of $F(f, \theta)$ as an array of frequency–direction bins of finite width Δf, $\Delta\theta$ collects a continuum of wave components traveling at slightly different group velocities in a single bin. The difference between the group velocities of the individual components and the mean bin group velocity results in a separation of the individual components of the bin relative to the mean bin position. The separation scale increases linearly in time. A wave packet which was initially concentrated at a single grid point becomes extended after a time t by a distance

$$\xi = \frac{dv}{df} \cdot \Delta f \cdot t \tag{B.1}$$

in the propagation direction and by

$$\eta = \Delta\theta \cdot v \cdot t \tag{B.2}$$

in the lateral direction.

A consistent numerical propagation scheme should reproduce these theoretical linear dispersion rates. Figure B.1 shows the spatial and spectral distribution at time t which would be predicted by a discrete spectral model with an ideally adapted propagation operator for the case of a wave field which is initially concentrated at a single grid point $x = y = 0$. The form of the spectrum is irrelevant for the discussion, but for simplicity we assume $F(f,\theta) = \text{constant}$ in the quadrant $0 \leq \theta \leq \pi/2$ and the frequency interval $f_{\min} < f < f_{\max}$. After the time t the separate spectral components have propagated into a set of separated but adjacent spatial "boxes" (denoted by dashed lines). The spectral energy density of a given spectral bin is inversely proportional to the area of its associated box, which increases quadratically with time. The integrated energy $E = \int F(f,\theta)\, df\, d\theta$ is quasi-continuous with respect to space and time, and also decreases quadratically with time along the wave propagation rays. Within the constraints of discretization, the model is seen to reproduce the properties of the exact solution

$$F(f,\theta;\mathbf{x},t) = \delta(\mathbf{x} - \mathbf{v}t)\, E_0$$

$$(f_{\min} \leq f \leq f_{\max},\ 0 \leq \theta \leq \pi/2,\ E_0 = \text{constant}) \tag{B.3}$$

$$E(\mathbf{x},t) = \int F(f,\theta;\mathbf{x},t)\, df\, d\theta = E_0 \left(\frac{dv}{df} \cdot vt^2 \right)^{-1} \tag{B.4}$$

In contrast, a "perfect" advection scheme without dispersion would yield the "garden sprinkler" distribution of discrete energy "droplets" indicated in Fig. B.1 by the solid boxes. Between the droplets, the wave energy is identically zero. Within the droplets, the energy is contained in a single frequency–direction bin, as before, but in contrast to the dispersive case, the energy density of the bin remains constant, rather than decreasing with the square of the propagation time. As time increases, the area between the droplets increases as t^2 relative to the constant area of the droplets, and the wave field becomes more and more spotty.

All existing discrete wave models which use forward propagation schemes (as opposed to ray back-tracing methods) suffer from the spottiness problem. Although numerical advection schemes provide some form of numerical dispersion, the dispersion length scale normally increases more slowly than t (typically as $t^{1/2}$), so that the dispersion rate cannot be

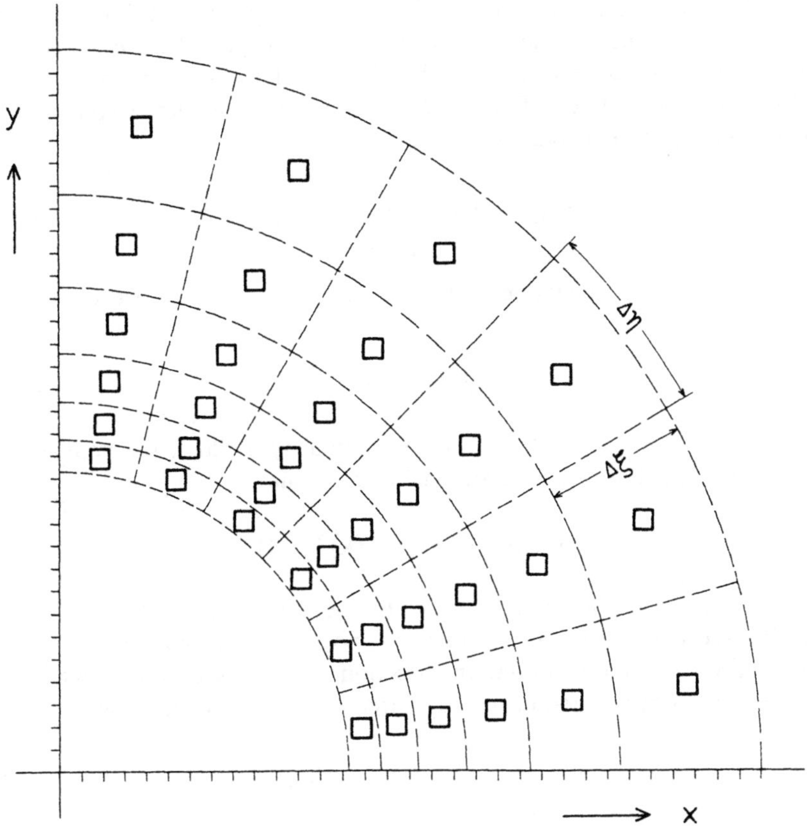

Fig. B.1. Areas covered by individual spectral bins at time t for a wave field initially concentrated at the origin $x = y = 0$ in the case of (1) dispersive propagation in accordance with equations (B.1) and (B.2) (dashed boxes), and (2) nondispersive propagation (solid boxes —"droplets"—of the same size as the spatial grid resolution box). In case (1) the wave field is quasi-continuous; in case (2) the wave energy is concentrated within the "droplets" and is zero in the intermediate areas.

adjusted to reproduce (B.1) and (B.2). Thus the wave field always becomes spotty asymptotically.

It can be shown that a propagation operator with the desired linear dispersion properties (B.1) and (B.2) must be nonlocal and therefore, in contrast to normal advection operators, cannot be readily approximated by a low-order local finite difference expression. For simplicity we investigate the properties of such an operator in a continuous spatial representation.

We assume that at time $t_0 = 0$ the initial spatial distribution of the energy in a particular frequency bin is given by $F_0(\mathbf{x})$, where $F_0(\mathbf{x})$ is some narrow, approximately δ-function distribution of width ξ_0, η_0 in the x and y directions, respectively. The spatial distribution at the time step $t_1 = \Delta t$ is then given by the convolution

$$F_1(\mathbf{x}) = \int P(\mathbf{x}-\mathbf{x}')\, F_0(\mathbf{x}')\, d\mathbf{x}' \equiv PF \tag{B.5}$$

where P is the propagation operator. After n time steps we obtain the distribution

$$F_n(\mathbf{x}) = P^n F \tag{B.6}$$

In terms of the Fourier transforms (denoted by a tilde), the convolution product P^n in (B.6) becomes an algebraic product,

$$\tilde{F}_n(\mathbf{k}) = (2\pi)^{2n} \cdot \tilde{P}(\mathbf{k})^n \cdot \tilde{F}_0(\mathbf{k}) \tag{B.7}$$

We attempt now to construct an operator P such that the initial distribution is displaced at the time $t_n = n\,\Delta t$ by the distance $d_n = vn\,\Delta t$ in the x direction (the propagation direction of the wave group), while the distribution is broadened in the x, y directions to the width scales

$$\xi_n = \frac{dv}{df} \cdot \Delta f \cdot (n+1)\,\Delta t \tag{B.8}$$

$$\eta_n = \Delta\theta \cdot v \cdot (n+1)\,\Delta t \tag{B.9}$$

respectively, in accordance with (B.1) and (B.2):

$$P^n \cdot F_0(x, y) = F_0\left[(x - d_n)\xi_0/\xi_n,\, y\eta_0/\eta_n\right] \tag{B.10}$$

We assume that the initial widths ξ_0, η_0 are consistent with the relations (B.8) and (B.9). The initial width normalization is chosen for convenience in order to yield a strictly linear dispersion rate from the outset. This is not necessary, however, and the following analysis can be applied for an arbitrary narrow initial distribution in the asymptotic sense for large n.

We do not yet prescribe the form of the distributions F_n. At this point we require only that the distributions for different n be self-similar, and that the mean positions d_n of the distributions and their characteristic longitudinal and lateral width parameters ξ_n, η_n vary linearly in time t in accordance with the requirements of a discrete spectral model. This is clearly a permissible generalization of the special box-car spectral and spatial windows assumed for simplicity in the discussion of Fig. B.1.

The Fourier transform of equation (B.10) yields the condition

$$(2\pi)^{2n}\tilde{P}^n(k_x,k_y)\tilde{F}_0(k_x,k_y)=\exp(-ik_xd_n)\tilde{F}_0(k_x\xi_n/\xi_0,k_y\eta_n/\eta_0) \tag{B.11}$$

or

$$n\left[2\ln(2\pi)+(ik_xv\Delta t)+\ln\tilde{P}(k_x,k_y)\right]=\ln\tilde{F}_0\left[k_x(n+1),k_y(n+1)\right]-\ln\tilde{F}_0(k_x,k_y) \tag{B.12}$$

Comparing the dependence on n of the left- and right-hand sides of (B.12), it follows that $\ln\tilde{F}_0$ must be a linear function of its arguments. This implies that $\ln\tilde{P}$ must then also be a linear function, and one obtains the general solutions

$$\tilde{P}(k_x,k_y)=(2\pi)^{-2}\exp(ik_xv\Delta t-\xi_0'k_x-\eta_0'k_y) \tag{B.13}$$

$$\tilde{F}_0(k_x,k_y)=A\exp(-\xi_0'k_x-\eta_0'k_y) \tag{B.14}$$

where A, ξ_0', and η_0' are arbitrary constants. If we impose, however, the condition $\tilde{F}_0\to 0$ for $|k_x|,|k_y|\to\infty$, and the requirement that the scale lengths of F_0 are given by ξ_0,η_0, we must choose $\xi_0'=\pm\xi_0$ for $k_x\gtrless 0$, $\eta_0'=\pm\eta_0$ for $k_y\gtrless 0$, and (B.13) and (B.14) become

$$\tilde{P}(k_x,k_y)=(2\pi)^{-2}\exp(ik_xv\Delta t-\xi_0|k_x|-\eta_0|k_y|) \tag{B.15}$$

$$F_{\tilde{0}}(k_x,k_y)=A\exp(-\xi_0|k_x|-\eta_0|k_y|) \tag{B.16}$$

where

$$\xi_0=\frac{dv}{df}\Delta f\cdot\Delta t \tag{B.17}$$

$$\eta_0=\Delta\theta\, v\cdot\Delta t \tag{B.18}$$

Transforming back to the original functions, we obtain for the propagation operator P, the initial distribution F_0, and the distribution F_n after n

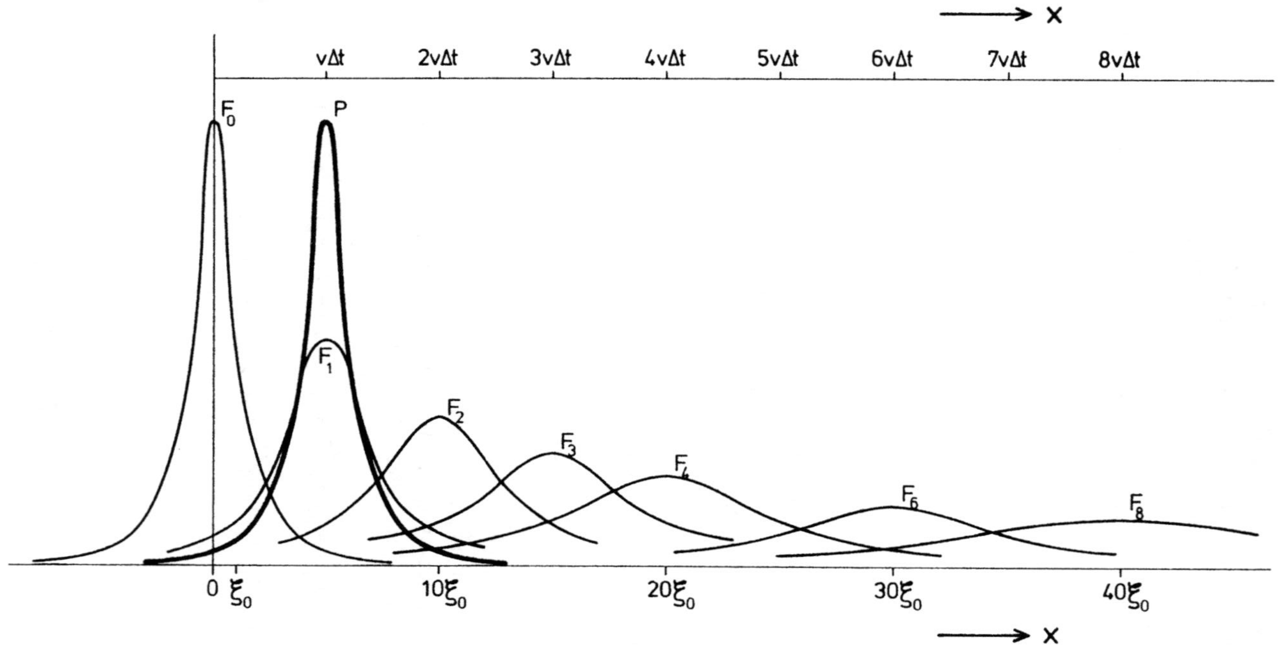

Fig. B.2. Propagation operator $P(x)$ and distribution function $F_n(x)$ for linear dispersion and advection appropriate for a discrete spectral model (one-dimensional x-dependent factors in the propagation direction; the lateral y-dependent factors are the same except for the missing linear displacement).

time steps, respectively:

$$P(x,y)=\frac{1}{(2\pi)^2}\frac{\xi_0}{\xi_0^2+(x-v\Delta t)^2}\cdot\frac{\eta_0}{\eta_0^2+y^2} \tag{B.19}$$

$$F_0(x,y)=\frac{A}{(2\pi)^2}\frac{\xi_0}{\xi_0^2+x^2}\cdot\frac{\eta_0}{\eta_0^2+y^2} \tag{B.20}$$

$$F_n(x,y)=\frac{A}{(2\pi)^2}\frac{\xi_0(n+1)}{(n+1)^2\xi_0^2+(x-nv\Delta t)^2}\cdot\frac{\eta_0(n+1)}{(n+1)^2\eta_0^2+y^2} \tag{B.21}$$

The functions are illustrated in Fig. B.2.

The appropriate propagation operator for a discrete spatial grid (or a one-dimensional characteristic propagation method) can be obtained from equation (B.19) by discretizing the integral representation (B.5).

References

Barnett, T. P. (1968): On the generation, dissipation and prediction of ocean wind waves. *J. Geophys. Res.* **73,** 513–530.

Barnett, T. P., and K. E. Kenyon (1975): Recent advances in the study of wind waves. *Rep. Prog. Phys.* **38,** 667–729.

Barnett, T. P., and J. C. Wilkerson (1967): On the generation of ocean wind waves as inferred from airborne measurements of fetch-limited spectra. *J. Mar. Res.* **25,** 292–328.

Donelan, M. A. (1981): A simple numerical model for wind and wave stress prediction. *Proc. Symp. Wave Dynamics and Radio Probing of the Ocean Surface*, Miami, Florida 1981.

Ewing, J. A. (1971): A numerical wave prediction method for the North Atlantic Ocean. *Dtsch. Hydrogr. Z.* **24,** 241–261.

Gelci, R., and P. Chavy (1978): Seven years of numerical wave prediction with the DSA5 model. *Turbulent Fluxes through the Sea Surface, Wave Dynamics, and Prediction* (A. Favre and K. Hasselmann, eds.), Plenum Press, New York, 565–591.

Gelci, R., and E. Devillaz (1970): Le calcul numérique de l'état de la mer. *Houille Blanche* **25,** 117–131.

Gelci, R., H. Cazalé, and J. Vassal (1957): Prévision de la houle. La méthode des densités spectroangulaires. *Bull. Inform. Comité Central Oceanogr. D'Etude Côtes*, **9,** 416–435.

Hasselmann, K. (1960): Grundgleichungen der Seegangsvoraussage. *Schiffstechnik* **7,** 191–195.

Hasselmann, K. (1961): On the nonlinear energy transfer in a wave spectrum. *Proc. Conf. Ocean Wave Spectra*, Easton, Maryland, 1961, 191–200.

Hasselmann, K. (1962): On the nonlinear energy transfer in a gravity-wave spectrum. 1. General theory. *J. Fluid Mech.* **12,** 481–500.

Hasselmann, K. (1963a): On the nonlinear energy transfer in a gravity-wave spectrum. 2. Conservation theorems, wave–particle correspondence, irreversibility. *J. Fluid Mech.* **15,** 273–281.

Hasselmann, K. (1963b): On the nonlinear energy transfer in a gravity-wave spectrum. 3. Computation of the energy flux and swell–sea interaction for a Neumann spectrum. *J. Fluid Mech.* **15,** 385–398.

Hasselmann, K., T. P. Barnett, E. Bouws, H. Carlson, D. E. Cartwright, K. Enke, J. A. Ewing, H. Gienapp, D. E. Hasselmann, P. Kruseman, A. Meerburg, P. Müller, D. J. Olbers, K. Richter, W. Sell, and H. Walden (1973): Measurements of wind–wave growth and swell decay during the Joint North Sea Wave Project (JONSWAP). *Dtsch. Hydrogr. Z., Suppl. A.* **8**(12).

Hasselmann, K., D. B. Ross, P. Müller, and W. Sell (1976): A parametrical wave prediction model. *J. Phys. Oceanogr.* **6,** 201–228.

Hasselmann, S., K. Hasselmann, T. P. Barnett, and J. H. Allender (1985): Improved methods of computing and parameterizing the nonlinear energy transfer in a gravity wave spectrum (submitted for publication).

Haug, O. (1968): A numerical model for prediction of sea and swell. *Nor. Meteorol. Inst. Meteorol. Ann.* **5**(4), 139–161.

Holthuijsen, L. H. (1983): Observations of the directional distribution of ocean wave energy in fetch limited conditions. *J. Phys. Oceanogr.* **13,** 191–207.

Jeffreys, J. (1924): On the formation of waves by wind. *Proc. R. Soc. London Ser. A* **107,** 189.

Kahma, K. K. (1981): A study of the growth of the wave spectrum with fetch. *J. Phys. Oceanogr.* **11,** 1503–1515.

Kawai, S., K. Okada, and Y. Toba (1977): Field data support of three-seconds power law and $gu_* \sigma^{-4}$-spectral form for growing wind waves. *J. Oceanogr. Soc. Jpn.* **33**(3), 137–150.

Kitaigorodskii, S. A., and M. M. Zaslavsky (1974): A dynamical analysis of the drag coefficient at the sea surface. *Boundary Layer Meteorol.* **6,** 53–61.

Komen, G. J., S. Hasselmann, and K. Hasselmann (1984): On the existence of a fully developed windsea spectrum. *J. Phys. Oceanogr.* **14,** 1271–1285.

Lazanoff, S. M., and N. Stevenson (1975): An evaluation of a hemispherical operational wave spectra model. Technical Note 75-3, Fleet Numerical Weather Center, Monterey, California.

Liu, P. C., and D. B. Ross (1980): Airborne measurements of wave growth for stable and unstable atmospheres in Lake Michigan. *J. Phys. Oceanogr.* **10,** 1842–1853.

Miles, J. W. (1957): On the generation of surface waves by shear flows. *J. Fluid Mech.* **3,** 185–204.

Mitsuyasu, H. (1968a): On the growth of the spectrum of wind-generated waves. 1. *Rep. Res. Inst. Appl. Mech., Kyushu Univ.* **16,** 459–465.

Mitsuyasu, H. (1968b): A note on the nonlinear energy transfer in the spectrum of wind-generated waves. *Rep. Res. Inst. Appl. Mech., Kyushu Univ.* **16,** 251–264.

Mitsuyasu, H. (1969): On the growth of the spectrum of wind-generated waves. 2. *Rep. Res. Inst. Appl. Mech., Kyushu Univ.* **17,** 235–243.

Mitsuyasu, H., R. Nakayama, and T. Komori (1971): Observations of the wind and waves in Hakata Bay. *Rep. Res. Inst. Appl. Mech., Kyushu Univ.* **19,** 37–74.

Mitsuyasu, H., F. Tasai, T. Suhara, S. Mizuno, M. Ohkusu, T. Honda, and K. Rikiishi (1975): Observation of the directional spectrum of ocean waves using a cloverleaf buoy. *J. Phys. Oceanogr.* **5,** 750–760.

Mitsuyasu, H., F. Tasai, T. Suhara, S. Mizuno, M. Ohkusu, T. Honda, and K. Rikiishi (1980). Observation of the power spectrum of ocean waves using a cloverleaf buoy. *J. Phys. Oceanogr.* **10,** 286–296.

Phillips, O. M. (1957): On the generation of waves by turbulent wind. *J. Fluid Mech.* **2,** 417–445.

Phillips, O. M. (1958): The equilibrium range in the spectrum of wind-generated ocean waves. *J. Fluid. Mech.* **4,** 426–434.

Pierson, W. J., Jr., and L. Moskowitz (1964): A proposed spectral form for fully developed wind seas based on the similarity theory of S. A. Kitaigorodskii. *J. Geophys. Res.* **69,** 5181–5190.

Pierson, W. J., Jr., L. J. Tick, and L. Baer (1966): Computer based procedure for preparing global wave forecasts and wind field analysis capable of using wave data obtained from a spacecraft. *Proc. Sixth Symp. Naval Hydrodynamics*, Washington, D.C., 499 pp.

Resio, D. T., and C. L. Vincent (1979): A comparison of various numerical wave prediction techniques. *11th Annual Offshore Technology Conference*, Houston, Texas, April 30–May 3, 1979, 2471–2478.

Ross, D. B. (1976): A simplified model for forecasting hurricane generated waves (Abstract). *Bull. A.M.S.* **57**, 113.

Ross, D. B., and V. J. Cardone (1978): A comparison of parametric and spectral hurricane wave prediction products. *Turbulent Fluxes through the Sea Surface, Wave Dynamics, and Prediction* (A. Favre and K. Hasselmann, eds.), Plenum Press, New York, 647–665.

Sand, S. E., O. Brink-Kjaer, and J. B. Nielsen (1982): Directional numerical models as design basis. *Proc. Conf. Wind and Wave Directionality*, Paris, 1981.

Sanders, J. W. (1976): A growth-stage scaling model for the wind-driven sea. *Dtsch. Hydrogr. Z.* **29**, 136–161.

Sanders, J. W., and J. Bruinsma (1981): Modeling wind driven sea in shallow water. *Proc. Symp. Wave Dynamics and Radio Probing of the Ocean Surface*, Miami, Florida 1981.

Snyder, R. L., F. W. Dobson, J. A. Elliott, and R. B. Long (1981): Array measurements of atmospheric pressure fluctuations above surface gravity waves. *J. Fluid Mech.* **102**, 1–59.

SWAMP group (1982): The sea wave modeling project, part 2: Compilation of results. Report no. 161, KNMI, de Bilt, The Netherlands.

Toba, Y. (1972): Local balance in the air–sea boundary processes. I. On the growth processes of wind waves. *J. Oceanogr. Soc. Jpn.* **28**, 109–120.

Toba, Y. (1973): Local balance in the air–sea boundary processes. III. On the spectrum of wind waves. *J. Oceanogr. Soc. Jpn.* **29**, 209–220.

Toba, Y. (1978): Stochastic form of the growth of wind waves in a single-parameter representation with physical implications. *J. Phys. Oceanogr.* **8**, 494–507.

Wilson, B. W. (1965): Numerical prediction of ocean waves in the North Atlantic for December 1959. *Dtsch. Hydrogr. Z.* **18**, 114–130.

II

The SWAMP Models

Chapter 15

The MRI Wave Model

Takeshi Uji

15.1. INTRODUCTION

The MRI wave model was developed at the Meteorological Research Institute (MRI) for routine wave forecasting at the Japan Meteorological Agency (JMA). It is a slightly modified version of the model discussed by Uji and Isozaki (1972), Isozaki and Uji (1983), and Uji (1975). Wave prediction in the northwestern Pacific Ocean is now in operation using this model with $Dx = 381$ km and $Dt = 6$ h. The wind field is calculated using Cardone's (1969) wind model. The model requires the atmospheric pressure and temperature as well as the temperature at the sea surface. Before the start of the prediction, objectively analyzed atmospheric pressure and temperature are input. The future atmospheric pressure and temperature are prognostically obtained using JMA's Four-Layer Northern Hemisphere Model. Monthly mean of sea surface temperature based on data covering a period of 15 years are used as sea surface temperature.

15.2. FUNDAMENTAL EQUATIONS

This model is based on the two-dimensional (2d) concept of ocean waves. The energy balance equation,

$$\frac{DS(f,\theta,t,x)}{Dt} = F(f,\theta,t,x) \tag{15.1}$$

Takeshi Uji • Meteorological Research Institute, Ibarakiken 305, Japan.

is the fundamental equation of our model, where S is the 2d energy spectrum defined as a function of the frequency f, the wave direction θ, the time t, and the position x. The source function F represents processes that are adding energy to, or subtracting energy from, the spectrum. Three stages of the sea state are considered, and the source function F is assumed according to each stage. First, during the growing stage, the energy balance equation of Inoue (1967) is modified to a 2d form,

$$\frac{\partial S}{\partial t} = -C_g(f,\theta)\cdot\nabla S + \left(A(f,u) + B(f,u_*)\cdot S\right)\left[1 - \left(\frac{S}{S_\infty\cdot\Gamma(\psi)}\right)^2\right]\Gamma(\psi) \tag{15.2}$$

for $\psi \leq 90°$ and $\sqrt{2}\,\Gamma S_\infty \geq S$, where C_g is the group velocity of the wave component, u the wind speed at t and x, u_* the friction velocity, and ψ the angle between θ and the wind direction. The term $(A + B\cdot S)\Gamma(\psi)$ represents energy supplied to the spectrum by the effect of local winds and the term $-(A + B\cdot S)\Gamma(\psi)[S/(S_\infty\cdot\Gamma(\psi))]^2$ represents energy subtraction from the spectrum by wave breaking. S_∞ is the fully developed one-dimensional spectrum proposed by Pierson and Moskowitz (1964) and $\Gamma(\psi)$ is the angular distribution of 2d spectral density and is assumed to be proportional to $\cos^2\psi$. The numerical values of A and B are calculated using the formulas

$$A(f,u) = \int_{-\pi/2}^{\pi/2} \frac{3.54\cdot 10^{-11}\omega^{5.25}u^{2.25}}{\left[\frac{1}{4}\left(\frac{\omega}{u}\right)^2 + (k\cdot\sin\theta)^2\right]\left[\frac{1}{9}\left(\frac{\omega}{u}\right)^{2.5} + \left(k\cdot\cos\theta - \frac{\omega}{u}\right)^2\right]}\,d\theta$$

and

$$B(f,u_*) = 5\exp\{-7000[(u_*/c) - 0.031]^2\} + 2612(u_*/c)^2\exp\left[-0.0004(c/u_*)^2\right]\cdot f$$

where $\omega = 2\pi f$, $k = \omega^2/9.8$, c is the phase velocity of the wave component, f is frequency in Hz, and the spectrum S is measured in units of $m^2\cdot s$. Equation (15.2) gives a rate of spectrum change per hour. These two formulas, proposed by Inoue (1967), were obtained as a result of analysis of wind and wave data in the early growing stage of wind waves, assuming mechanisms proposed by Phillips (1957) and Miles (1957).

The energy-subtracting term, $-(A + B\cdot S)\Gamma(\psi)[S/(S_\infty\cdot\Gamma(\psi))]^2$, is introduced according to the qualitative hypothesis that a fully developed sea state should exist as a stable equilibrium. Therefore the range of application of equation (15.2) is $S < S_\infty$, or at most it extends to $S \simeq S_\infty$. An alternative equation for the state $S > S_\infty$ is difficult to formulate because of poor information on this state, especially concerning wave breaking. For practical purposes, the upper limit of the range of application of equation (15.2) is extended from S_∞ to $1.414S_\infty$, the qualitatively rational limit of equation (15.2) itself. Above the new limit, equation (15.2) has a larger damping effect on wind waves than equation (15.3) has for opposing winds.

In the second stage, the decaying stage, only frictional dissipation is assumed to play a role. The governing equation is

$$\frac{\partial S}{\partial t} = -C_g \cdot \nabla S - D \cdot f^4 S \tag{15.3}$$

where $\psi = 90°$, $\sqrt{2}\, S_\infty \Gamma < S$, and D is a numerical constant. In our old numerical experiments (Isozaki and Uji, 1973; Uji, 1975), the numerical value of D was 81, when f is in hertz and time t is in hours. This value of D was adopted so that the wave components whose period is shorter than 3 s would die out within 1 h if there was no energy supply from wind. From test runs of the wave forecast at JMA over almost half a year, we found that a D value of 81 underestimates wave heights in the area where a storm has passed, in spite of a good estimation of wave height within the storm area. Hence, a D value of 1 is adopted, and this value causes very little energy dissipation in stage 2. Although equation (15.3) indicates that a swell travels on a calm sea in the same way as it travels through a sea under the action of local winds, the nonlinear energy transfer between swell and wind waves may accelerate the energy damping of the swell. Thus the nonlinear swell–wind wave interaction is omitted in equation (15.3). In any event, energy dissipation is supposed to be very small when the wind has lulled, i.e., $S > 1.414S_\infty$.

The third stage is for opposing winds. In this stage, it is assumed that the winds produce reverse effect to wave generation. This assumption is supported experimentally by Mitsuyasu and Mizuno (1971). The equation for this stage is

$$\frac{\partial S}{\partial t} = -C_g \cdot \nabla S - \left(B(f, U_*)\Gamma(\psi) + D \cdot f^4\right) S \tag{15.4}$$

where $\psi > 90°$ and B and D are the same as those in the stage 1 equation and the stage 2 equation, respectively.

It is apparent that our model neglects the effects of wave–wave interactions. It is also seen from equations (15.2)–(15.4) that three kinds of data, i.e., wind direction, wind velocity, and friction velocity, are required as input data for the operation of the present wave model.

15.3. BOUNDARY CONDITIONS

Absence of information about waves beyond the domain of calculation does not allow the calculation of propagation of wave components traveling inward at the boundary. Accordingly, a boundary condition for the advection of incoming wave energy components at the boundary should be introduced.

Along the "closed boundary" where a domain of calculation is bounded by land, S is simply set as zero. If the boundary of the area is sea or an "open boundary," a special condition is used. For example, when a big typhoon comes into the calculation area, very strong wind blows inward on some part of the boundary. One extreme condition for this situation is to assume that the next outward grid points are a rigid boundary. The other extreme is to suppose that the sea extends infinitely with the wind the same as that at the boundary, i.e., $\partial S/\partial m = 0$, where m is the position along the travel direction of a wave component. The former condition leads to very small wave height and the latter leads to unusually large wave energy input along the boundary. Considering this situation, the wave components traveling inward at the open boundary are assumed to be governed by the following equations.

For stage 1, the advection term is dropped and the fully developed energy level is suppressed as

$$
\begin{aligned}
\partial S/\partial t &= (A + B\cdot S)\left\{1 - [S/(S_\infty \Gamma)]^2\right\} & f > 0.1 \\
\partial S/\partial t &= (A + B\cdot S)\left\{1 - [S/(10\cdot f\cdot S_\infty \Gamma)]^2\right\} & f \le 0.1
\end{aligned}
\qquad (15.5)
$$

where the frequency f is measured in hertz.

For stage 2, equation (15.5) is assumed to be applied in order to reduce the wave energy rapidly with decrease of wind velocity.

For stage three, $S = 0$.

The factor $10\cdot f$ in equation (15.5) is experimentally determined in order that the wave height distribution along the open boundary when the center of a typhoon is just crossing into the calculation area be almost the same as the wave height distribution along a line whose relative position to the typhoon, traveling near the center of the calculation area, is the same as the position of the boundary relative to the incoming typhoon at the boundary (Uji, 1975).

15.4. SCHEME OF THE CALCULATION

Cartesian coordinates are employed and deep-sea waves are assumed. The wave energy is divided into 16 directional components clockwise from N to NNW and 22 frequency components from 0.04 to 0.25 Hz. The components have spectral width of 22.5° and 0.01 Hz for direction and frequency. Accordingly, $S(f_i, \theta_j, t, x)$ is the representative value of the frequency band from $(f_i + 0.005)$ Hz to $(f - 0.005)$ Hz and of the band of the travel direction from $(\theta_j + 11.25)°$ to $(\theta_j - 11.25)°$.

15.4.1. Propagation

The scheme for calculating propagation of wave energy is designed so that each wave energy component moves like a particle with a constant velocity C_g unless energy dispersion owing to the spectral width of the component is taken into account. Since any wave component does have spectral width, it is more reasonable to assume that a wave component spreads its own energy during the propagation than to assume that no spreading is occurring. Thus, the following two procedures are introduced.

15.4.1.1. Lateral Spreading of Wave Energy

Consider a wave component that has reached the point P(I, J) along the direction θ at the time t_1 in Fig. 15.1. First we need know the wave field along the direction θ at the time $t_1 - Dt$. In Fig. 15.1, the energy density S_x

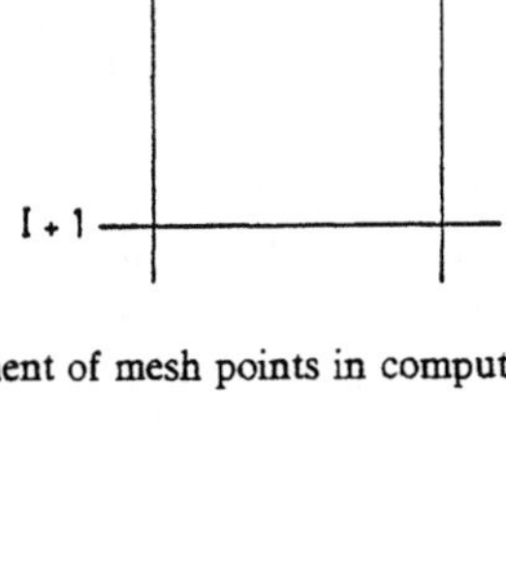

Fig. 15.1. Arrangement of mesh points in computation of wave propagation.

at the point X is

$$S_x = [bS_{I-1,J+1} + (a-b)S_{I-1,J}]/a \tag{15.6}$$

Then, the representative value of energy density along the direction θ is

$$S_n = \frac{8}{\pi}\int_{\theta-\pi/16}^{\theta+\pi/16} S_x\, d\theta \tag{15.7}$$

Using equation (15.6), equation (15.7) is reduced to the equation

$$S_n = C_1 \cdot S_{I+\alpha, J+\alpha'} + C_2 \cdot S_{I+\beta, J+\beta'} + C_2 \cdot S_{I+\gamma, J+\gamma'} \tag{15.8}$$

where C_1, C_2, C_3, α, α', β, β', γ, and γ' are constants which are determined depending on the travel direction of the wave component. The gradient of the energy density $G_{I,J}$ along the travel direction is calculated by the equation

$$G_{I,J} = \frac{S_n - S_{I,J}}{a}\cos(Q)$$

where Q is the smallest angle between the θ direction and the axes. Accordingly, the energy density of waves at the point P(I, J) at the time t_1 is

$$(S_{I,J})_{t=t_1} = (S_{I,J})_{t=t_1-Dt} + (G_{I,J})_{t=t_1-Dt} C_g \cdot Dt \tag{15.9}$$

This technique is quite equivalent to the finite difference method, and seems to give a reasonable lateral spreading of wave energy. The actual calculation of this equation, however, results in too much dispersion of wave energy along the travel direction, because the wave front must progress one grid interval at every time step and the rear of the wave remains at the original point. This false dispersion is prevented by the method described in the next section.

15.4.1.2. Dispersion and Propagation along Travel Direction

A component $S(f_i, \theta_j, t, x)$ is a representative value of the frequency band $(f_i + 0.005)$ Hz to $(f_i - 0.005)$ Hz. As the group velocity varies from $g/[4\pi(f_i + 0.005)]$ to $g/[4\pi(f_i - 0.005)]$ in the frequency band, the wave energy will disperse during the propagation along the travel direction, where $g = 9.8\ \mathrm{m/s^2}$. Consider successive points $Y_1, Y_2, \ldots$ at regular interval Dy on the path of a directional wave component as shown in Fig. 15.2. The

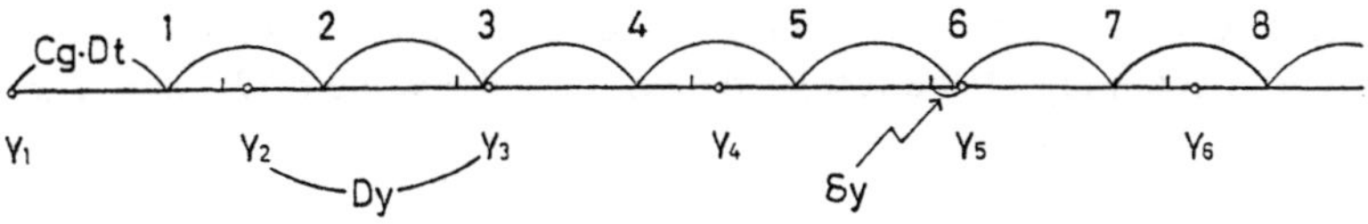

Fig. 15.2. Schematic expression of the procedures of longitudinal wave dispersion and propagation.

interval, Dy, of successive points is equal to the distance PX in Fig. 15.1. Let us put a small distance δy at each point on the path. Wave components progress over the distance $|C_g| \cdot Dt$ every one time step to the points which are classified into the following three cases:

Case 1. The point does not reach the position $Y_{I+1} - \delta y$.
Case 2. The point drops into the distance δy.
Case 3. The point goes beyond the point Y_{I+1}.

In Fig. 15.2 the time steps 1, 4, and 7 fall into case 1, the time steps 3 and 6 are for Case 2, and the time steps 2, 5, and 8 belong to Case 3. In case 1, the wave energy remains at the point Y_I. In case 2, the wave energy is divided into two parts, namely, half of S_Y goes into the point Y_{I+1} and the other half remains at the point Y_I. In case 3, S_Y is transferred to the point Y_{I+1}. For simplicity, the notation S_Y is used, but practically, S_Y corresponds to S_n of the previous section. Therefore the propagation, the lateral spreading, and the spreading along the travel direction of wave energy are calculated at the same time.

From the condition necessary for computational stability, the time step interval Dt and grid size Dx must be chosen to satisfy the relation

$$|C_g| \cdot Dt \leq Dx \qquad \text{i.e., Courant number } \leq 1$$

Apparently in our model this relationship is implicitly satisfied by the above-mentioned procedures. Hence, we can use any Dt independent of Dx.

15.4.2. Wind Effect

The wind effect is calculated after the calculation of wave propagation. By use of the limiting condition of S_∞, the calculation of wind effect can be executed under any value of Dt with a simple forward difference approximation without computational instability. (In the case of opposing wind, of course, S_∞ is assumed to be zero.) This is mainly due to the fact that the existence of equilibrium state S_∞ forbids the increase (or decrease) of wave energy density beyond S_∞ as time goes on, under any value of Dt. Needless to say, the accuracy of the results depends on the value of Dt and Dx.

15.5. REMARKS CONCERNING THE INPUT DATA SET

As already mentioned in the last part of Section 15.2, the wind direction, the wind velocity, and the friction velocity are required as input data. But the friction velocity u_* cannot always be obtained easily on an operational basis. Hence, we considered two models. One requires three kinds of data and the other requires only wind direction and wind velocity. In the latter model, the friction velocity u_* is converted into wind velocity by means of the relation established by Kuznetsov (1970):

$$10^2 u_* = 2.67u + 3.15 \qquad u \leq 11.14 \text{ m/s}$$

$$10^2 u_* = 12.14u - 105.0 \qquad u > 11.14 \text{ m/s}$$

In Case VI of the present intercomparison study, only the wind direction and wind velocity are given, so we had to use the latter model. For consistency through a series of experiments, the latter model is used through all of the cases, disregarding the given friction velocity. This choice cannot affect our results seriously because in our model the wind velocity data are much more important than the friction velocity.

Inoue's (1967) model, the basis for our model, was adjusted to give a wind velocity at 19.5 m above the sea surface, and so does our model. The wind velocity u_{10} was not converted to $u_{19.5}$ in all of the present experiments, i.e., $u_{19.5} = 20$ m/s is assumed. Further information about the stability of the boundary layer of the atmosphere is required to convert u_{10} to $u_{19.5}$. The value $u_{19.5} = 21.4$ m/s is obtained under the assumption of a neutral atmosphere. Our model uses the wind velocity given to an accuracy of the order of 1 m/s. Thus, even if a neutral condition is assumed, the conditions of the numerical experiment cannot be accurately adjusted to the assigned condition $E_{PM} = 6.0552$ m^2. Consequently the value $u_{19.5} = 20$ m/s is chosen as an adequate approximation. Although that value of wind velocity may cause some underestimation of the absolute value of wave energy, the characteristics of our model are nevertheless well demonstrated by the experiments, since in most case studies the results are normalized to remove the influences of different effective wind inputs.

15.6. DISCUSSION

An extended version of the final report on the Wave Model Intercomparison Test will include about 60 diagrams from our model, showing clearly its detailed characteristics. Here, the results are simply outlined.

The spectral form at an early growing stage of wind waves cannot be realized successfully by our growth term $A + B \cdot S$. We consider this to be the weakest point of our model. The peak frequency or mean frequency does not naturally come out very well at this stage. In spite of these points, the total energy grows reasonably with both time and fetch, because our numerical scheme for the advection term is reliable. The reliability of the energy propagation is one of the strong points of our model. The most remarkable point is that our model always gives reasonable wave heights for any wind field. This point is clearly shown by the wave height distribution of Case VI in the extended version.

The weakness mentioned above in the early growing stage does not seem to affect the wave height, because any serious discrepancy of the wave spectral form appears only at an early growth stage of wave generation. In that stage, the energy of wind waves is so small, e.g., less than 20% of the fully developed energy, that the distorted spectral form does not affect the wave height distribution.

ACKNOWLEDGMENTS. The author is indebted to Profs. Y. Toba and H. Mitsuyasu for introducing him to the Wave Model Intercomparison Study and for encouraging him to participate in the Symposium, and would also like to acknowledge the helpful advice of Prof. Toba. He would also like to express his thanks to all of those who have given him a chance to participate in the Symposium, especially to Dr. R. B. Long and Prof. K. Hasselmann.

REFERENCES

Cardone, V. J. (1969): Specification of the wind distribution in the marine boundary layer for wave forecasting. Report GSL-TR69-1, New York University School of Engineering and Science, 131 pp.

Inoue, T. (1967): On the growth of the spectrum of a wind generated sea according to a modified Miles–Phillips mechanism and its application to forecasting. Report TR-67-5, New York University Geophysical Sciences Laboratory, 74 pp.

Isozaki, I., and T. Uji (1973): Numerical prediction of ocean waves. *Pap. Meteorol. Geophys.* **24,** 207–232.

Kuznetsov, O. A. (1970): Results of an experimental investigation of air flow above the sea surface. *Izv., Atmos. Ocean. Phys.* [English edition] **6,** 469–472.

Miles, J. W. (1957): On the generation of surface waves by shear flows. *J. Fluid Mech.* **3**, 185–204.

Mitsuyasu, H., and S. Mizuno (1971): Experimental study of wave decay caused by opposing winds [in Japanese]. *Proc. 18th Conf. Coastal Engineering in Japan*, 37–42.

Phillips, O. M. (1957): On the generation of waves by turbulent wind. *J. Fluid Mech.* **2**, 417–445.

Pierson, W. J., and L. I. Moskowitz (1964): A proposed spectral form for fully developed seas based on the similarity theory of S. A. Kitaigorodskii. *J. Geophys. Res.* **69**, 5181–5190.

Uji, T. (1975): Numerical estimation of the sea wave in a typhoon area. *Pap. Meteorol. Geophys.* **26**, 199–217.

Uji, T., and I. Isozaki (1972): The calculation of wave propagation in the numerical prediction of ocean waves. *Pap. Meteorol. Geophys.* **23**, 347–359.

Chapter 16

The VENICE Model for Wind Wave Prediction

Luigi Cavaleri and Luciana Bertotti

16.1. THE MODEL

The VENICE model has been described by Cavaleri and Malanotte Rizzoli (1981). It is a physical model, based on the ray technique, suitable also for shallow-water applications. We will give here only a short description of the model, particularly the part concerned with shallow-water processes. These are not of interest in the actual study because of its deep-water assumption. Readers interested in the details of the model are referred to the paper quoted.

Given a bottom topography and a reference point A, we consider a wave crest of a given frequency f, passing through the point and moving in a specified direction. The path of the wave, or better, the ray normal to the wave crest, is established by the refraction laws (Collins, 1972)

$$\left.\begin{aligned} \frac{dx_i}{dt} &= v_i = \frac{\partial \sigma}{\partial k_i} \\ \frac{dk_i}{dt} &= -\frac{\partial \sigma}{\partial x_i} \end{aligned}\right\} \quad i=1,2 \tag{16.1}$$

where $x_i = (x_1, x_2)$ are the two coordinates in the plane, $k_i = (k_1, k_2)$ is the wave number, $\sigma = 2\pi f$ is the circular frequency, $v_i = (v_1, v_2)$ is the group velocity, and t is the time.

Luigi Cavaleri • Istituto per lo Studio della Dinamica delle Grandi Masse, Consiglio Nazionale delle Richerche, 30125 Venice, Italy. **Luciana Bertotti** • Associazione Studio Correnti Onde, 35100 Padua, Italy.

The ray, or characteristic, is also the path followed by the corresponding energy packet (frequency f, direction ϕ) proceeding on the ray with the group velocity v. Through a suitable discretization in frequency and direction, we can then through (16.1) establish the paths followed by all the energy packets approaching the target point A. If at a given time t_0 the value of energy $E\ (f,\phi)$ at A is known for each single characteristic, the corresponding two-dimensional (2d) matrix represents the wave directional spectrum at time t_0 at point A. Circular integration all over the directions provides the one-dimensional spectrum

$$E(f)=\int_0^{2\pi}E(f,\phi)\,d\phi$$

In a field of depth $h(x, y)$, σ and k are related by the dispersion relationship

$$\sigma^2=gk\tanh kh$$

As equations (16.1) are integrated in wavenumber space, the wave height variations with refraction and shoaling are automatically taken into account. The resulting spectral values are then transformed into (f,ϕ) space through the expression

$$E(k_1,k_2)=\frac{v}{2\pi k}E(f,\phi)$$

(Collins, 1972), k being the wavenumber modulus.

Energy approaching the target point along a certain ray is evaluated by the integration of the energy balance equation

$$\frac{\partial E}{\partial t}=\mathbf{v}\cdot\nabla E+S \tag{16.2}$$

proposed by Gelci *et al.* (1957), where S expresses the local algebraic energy input. Equation (16.2) expresses the time variation at a fixed point. If we focus our attention on the patch of energy moving along the ray with velocity v, the first term on the right-hand side of (16.2) vanishes, and we are left with the simple expression

$$\frac{\partial E}{\partial t}=S \tag{16.3}$$

valid along each single ray. The problem of evaluating the 2d spectrum

$E(f, \phi)$ is therefore reduced to that of specifying the source function S. This includes generative, conservative, and dissipative processes. Generation is represented by the Phillips and Miles processes.

Phillips. This derives from the turbulent pressure fluctuations associated with the wind (Phillips, 1957), and it leads to a linear energy growth

$$\frac{\partial E}{\partial t} = \alpha \qquad (16.4)$$

Care is necessary to avoid divergence of the expression for α in shallow water. The expression we chose is

$$\alpha(k) = \frac{80\rho_a^2\sigma}{\rho_w^2 g^2 k^2} c_d^2 u^4 \qquad (16.5)$$

It was given by Phillips (1966, pp. 123–125), and it is based on the measurements of Willmarth and Woolridge (1962) in a wind tunnel. In (5) ρ_a, ρ_w are air and water density, g is gravitational acceleration, $c_d = 0.0012$ is the drag coefficient, and u is the wind component along the ray direction.

Miles. The interaction of an already disturbed surface with the overlaying wind results in a feedback mechanism leading to an exponenetial energy growth (Miles, 1957) expressed by

$$\frac{\partial E}{\partial t} = \beta E \qquad (16.6)$$

For β we use the Barnett expression (1968),

$$\beta = 5sf\left(\frac{u\cos\delta}{c} - 0.90\right) \qquad (16.7)$$

where $s = \rho_a/\rho_w$ and δ is the angle between the wind and wave vectors. The 0.90 term allows for the experimental evidence of growing waves propagating faster than the wind.

The conservative processes of refraction and shoaling are automatically taken into account through the integration of (16.1).

The existence of nonlinear wave–wave interactions under certain resonance conditions was pointed out by Hasselmann (1962, 1963), and later proved during JONSWAP experiment (Hasselmann *et al.*, 1973). It is now believed that they play a key role in determining the shape of a growing spectrum, but their correct evaluation is made practically impossible by the large computer time that is required. Investigations have usually resorted to their parameterization (Barnett, 1968; Ewing, 1971; Resio and Vincent, 1977a, b), but this solution is problematic as the energy transfer among the different frequencies depends critically on the details of the spectrum. In any event, our model is basically uncoupled and linear, as each energy component is evaluated independently of the other ones, and the directional spectrum, necessary for the evaluation of the nonlinear interactions, is known only at the target position. Therefore, besides being ruled out by the practical complexity of their evaluation, there is not even in principle any way by which nonlinear interactions can be taken into account. This has dictated the choice of β of the Miles process [(16.6), (16.7)]; being deduced by direct fit to experimental data (Barnett and Wilkerson, 1967), it contains implicitly all information about the whole physical process we are trying to describe.

Dissipation acts in a double way: by producing a loss of energy during the growth phase (quantified by Hasselmann, 1974) and by establishing the limiting form of a saturated spectrum (Phillips, 1958). The original limiting form was given for a 1d spectrum. For the maximum energy along each single ray we have followed Barnett (1968) with the expressions

$$\begin{aligned} E_s(f,\delta) &= 0.073 g^2 \sigma^{-5} S(\delta) \\ S(\delta) &= \frac{8}{3\pi}\cos^4(\delta) \end{aligned} \tag{16.8}$$

thus assuming a $\cos^4$ spreading function.

The energy along each single ray is therefore allowed to grow according to (16.3), with a maximum value given by (16.8). Bottom friction, by far the most important bottom interaction process in the shallow areas where the model has been applied (Cavaleri and Malanotte Rizzoli, 1981), is a nonlinear process, and it was evaluated by using a "linearized" expression suggested by Hasselmann and Collins (1968). Bottom horizontal velocity was estimated assuming a parametric representation of $E(f)$ using the mean JONSWAP spectrum. Further discussion of this and other bottom processes (percolation, bottom elasticity, bottom scattering, mud waves) is avoided here because of the basic deep-water assumption of the present test.

Therefore, the only dissipative process we are left with is breaking. There is a certain degree of inconsistency in the contemporary use of the β of Barnett (1968) and the breaking expression of Hasselmann (1974). This β was established by a direct fit to experimental data, without taking breaking into account; this was therefore implicitly imbedded into (16.7). Nevertheless, we used it in the original version of the model to check its influence, and we found that, even if the process is a fundamental one from a physical point of view (Banner and Melville, 1976), it is quantitatively limited to a slight decrease of the energy growth rate, with little effect on the final energy value. We decided therefore to neglect it in the expression of the source function. Consequently the energy balance equation is now given by

$$\frac{\partial E}{\partial t} = \alpha + \beta E \tag{16.9}$$

with α and β given respectively by (16.5) and (16.7), and a limiting value expressed by (16.8).

Note that (16.9) can be analytically integrated with the solution

$$E(t) = -\frac{\alpha}{\beta} + \left(\frac{\alpha}{\beta} + E_0\right)\exp[\beta(t - t_0)] \tag{16.10}$$

where t_0, E_0 represent the initial time and energy values. For an initial null value of energy and writing $\Delta t = t - t_0$, (16.10) reduces to

$$E(t) = \frac{\alpha}{\beta}[\exp(\beta \cdot \Delta t) - 1] \tag{16.11}$$

α and β are established through (16.5) and (16.7) by the local (in time and space) wind component with respect to the actual wave direction, and by the wave frequency considered. If the wind field is uniform in space and time, for a given ray α and β are therefore constant, and equation (16.11) can consequently be integrated in a single step, according to the wind duration or fetch length, whichever of the two is revealed as the more restrictive.

In a complex wind field we will find rays running against the wind. While it is an implicit assumption that no energy is generated against the wind, any energy already existing on these rays will be attenuated. The corresponding physics is not yet completely understood, but the existence of the physical process is supported by intuition and experimental evidence. For attenuation against the wind we have used expression (16.6), with β given equally by (16.7). Note that, as $\cos\delta$ is now negative, the process is much more active than in the corresponding generation case, as if the wind had a much higher velocity.

A similar but different case arises when an energetic ray propagates along the wind ($\delta \leq 90°$), but with an energy larger than the corresponding saturation value given by (16.8). If the wave is decoupled from the wind, i.e., if $c > 1.3w$, there is no attenuation and the wave is propagated as swell. If not, we have used a time exponential decay $e^{-\gamma t}$ down to the actual saturation value. Hasselmann *et al.* (1976) estimate the response time of the nonlinear interactions for a strongly active wind to be of the order of 1 h, with higher values for decreasing ration w/c. Accordingly, the time constant of the exponential attenuation has been taken as

$$\gamma = \frac{(w/c)^2}{3600}\,\text{sec}^{-1}$$

Summarizing, for the actual test the model consists of a suitable discretization in frequency and direction, and a consequent number of rays (energy paths) starting from a given target point. For each f and ϕ, the ray is evaluated by the integration of (16.1). Because of the deep-water assumption, and consequent lack of any refraction and shoaling effect, the solution of (16.1) reduces to a simple advancement along the direction ϕ, linear in time, at group velocity v. Once the ray has been estimated, the associated energy is evaluated by the integration of (16.9), taking the saturated value into account. Circular integration of $E(f, \phi)$ provides the 1d spectrum $E(f)$. Integration in frequency provides the overall energy E, from which the significant wave height H_s is evaluated by means of

$$H_s = 4\sqrt{E}$$

Mean frequency $\bar{f}$ is estimated by a weighted average over frequency according to the energy distribution. Mean direction $\bar{\phi}$ is estimated by a weighted average over direction.

16.2. APPROACH TO THE DIFFERENT TEST CASES

Reference is now made to the general specifications of the intercomparison test. The approach is singularly described for each test, and a few typical results are reported.

16.2.1. Case I: Propagation

This case is not applicable to our model. The case was conceived to study the propagation algorithm on a grid and the associated dispersion of the initial energy distribution. But, working with the ray technique, we consider energy patches moving along their path with group velocity.

Focusing attention on the single patch, as is done with equation (16.3), no dispersion is possible and the test loses its meaning.

For the other six cases, dealing with actual generation, 16 frequencies [0.04–(0.005)–0.08, 0.09, 0.10, 0.11, 0.12, 0.15, 0.175, 0.20] and 24 directions (at 15° intervals) have been considered. 0° direction is to the north, and angles are considered as positive in the clockwise direction.

16.2.2. Case II: Fetch-Limited Growth

Consider a target point A (Fig. 16.1) at distance d from the coast. For a general direction ϕ the fetch AB is given by $L = d/\cos\phi$. For a frequency f with group velocity

$$v = \frac{g}{4\pi f}$$

the time necessary to run the ray is

$$t_1 = \frac{L}{v} = \frac{4\pi f L}{g} \tag{16.12}$$

If the spectrum at A is to be evaluated at time t after the wind has begun to blow, the shorter time between t and t_1 determines the actual integration time of (16.9). The resulting energy value $E(f, \phi)$ is then compared with

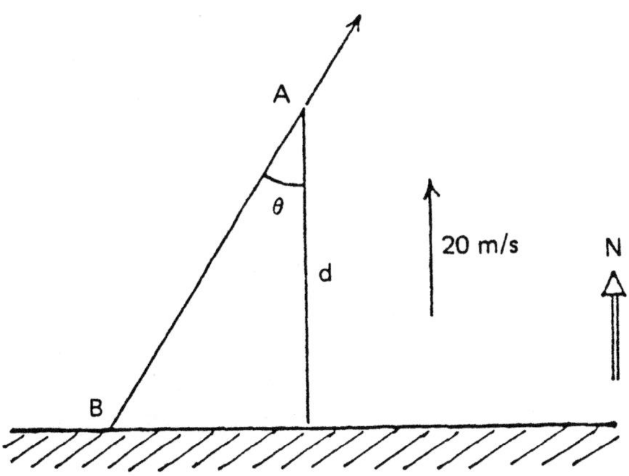

Fig. 16.1. Case II, fetch-limited growth. Because of the symmetry with respect to the north, six directions are sufficient for the definition of the 2d spectrum.

saturation conditions [(16.8)], and eventually reduced accordingly. Because of the fast exponential energy growth, it is convenient to check the exponent overflow. If $\beta \cdot \Delta t$ is very large, $E(f, \phi)$ is directly set equal to the saturated value. The evaluation proceeds from low to high frequencies. As saturation is first reached on higher frequencies, when for a given direction a frequency f_0 is saturated, similar conditions are automatically assumed for all $f > f_0$.

It is clear from Fig. 16.1 that symmetry exists with respect to the north. We have also implicitly assumed that no energy is generated against the wind. Hence evaluation can be limited to six directions, 0°,15°,30°, 45°,60°, and 75°. The wind being uniform in space and time, energy along each ray is estimated by a single application of (16.11). Symmetrical rays are assigned equal energy values. Fig 16.2 shows one example of a 2d spectrum and of the corresponding 1d one. Note that all the 2d spectra are normalized with respect to their own maximum value, reported in the figure, and shown as isolines at the 0.1 interval. While the general shape of the spectrum is clearly visible, any argument about its details is weakened by the influence of the interpolation procedure required to trace the isolines. The GPCP program was used for this purpose throughout the test, but its output is fairly sensitive to limited variations of the input parameters.

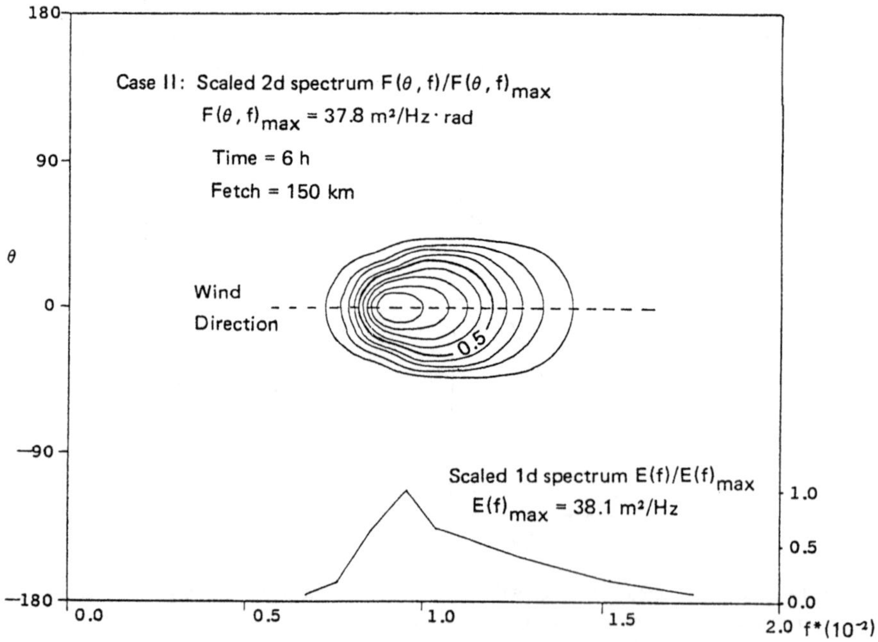

Fig. 16.2. Case II. Example of 2d and corresponding 1d spectrum.

Besides, while the real spectrum is smooth and the isolines have a natural continuity, the use of a discrete set of values for f and ϕ leads to the winding of isolines present in most of the 2d spectra. At this stage much of the information on the spectral details is lost.

16.2.3. Case III: Slanting Fetch, Limited Growth

This case (Fig. 16.3) is run similarly to Case II, with some variation in the estimate of the fetch length. To describe the general field the basin is covered with a net of 10×10 points, 75 km distant from each other. 2d and 1d spectra, H_s and f_p have been estimated at each single point. Scaled E and f_p values are evaluated with respect to the corresponding values of the Pierson–Moskowitz spectrum ($E_{PM} = 6.0552$ m^2, $f_{PM} = 0.6374$ Hz). Symmetry holds with respect to the wind direction, hence calculation is done for only half the basin, and the results are transferred to the symmetrical points. The 2d spectra are obviously symmetrical on the 45° line through the origin.

They are in any case narrower than in Case II, i.e., energy is more concentrated in the wind direction. This is due to the limited fetch length for the rays oblique with respect to the wind. For the same reason the 2d

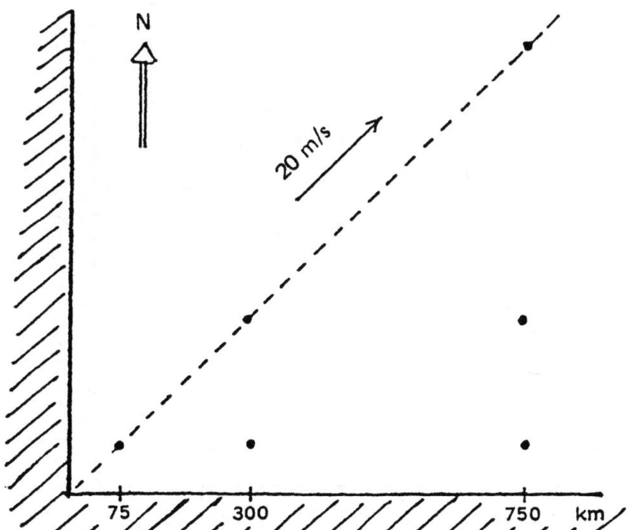

Fig. 16.3. Case III, slanting fetch, limited growth. The field has been covered with a net of 10×10 points. Grid step is 75 km. The six dots show the points chosen for special plots.

spectra become more and more asymmetrical while approaching the boundaries.

16.2.4. Case IV: Windy Half-Plane

The solution of this case is similar to that of the previous cases, with some modification in the estimate of the integration time. Steady state is considered, hence we can directly refer to the whole ray length. But, for the integration of (16.11), Δt must correspond only to the time necessary to run the ray section included in the windy half-plane (Fig. 16.4). The western and eastern boundaries now exist, and this poses a limit to the length of the lateral rays of many points. The basin has been covered with a net of 13×13 points, at a 75-km distance from each other, from whose values the various isolines have been interpolated. Nine extra points across the front (Fig. 16.4) have been considered for detailed analysis. Points close to the front but in the calm zone receive energy only along the rays coming from the windy zone. Hence, even if relatively energetic, their spectrum is fully asymmetrical.

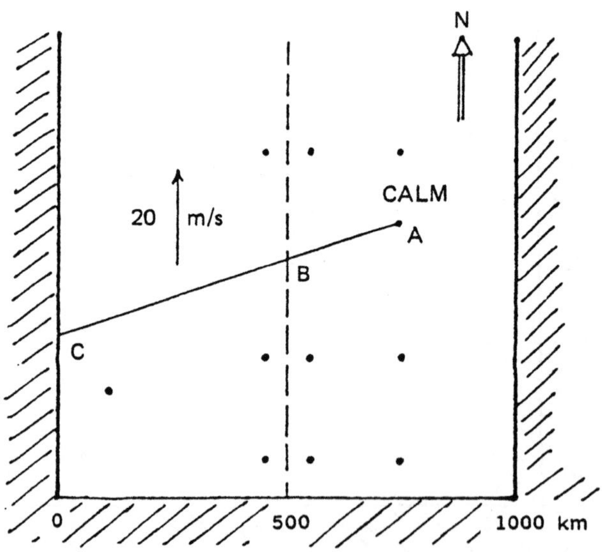

Fig. 16.4. Case IV, windy half-plane. The field has been covered with a net of 10×10 points. Grid step is 75 km. The ten dots show the points chosen for special plots. The actual fetch along the ray AC is equal to BC.

16.2.5. Case V: Frontal Passage

The basin is covered with a net of 13×13 points at a 75-km distance from each other, plus eight extra points close to and across the front. Calculations are singularly done for each point. Steady state must be considered, and the ray length up to the borders can be easily established for each point and direction (Fig. 16.5). Two areas with different wind are present, hence for each ray the estimate of the energy associated with a given f and ϕ is made by two sequential applications of (16.10).

The respective integration time $\Delta t = t - t_0$ is established by the corresponding ray section length (CB and BA in the example of Fig. 16.5). For points on the front or rays not crossing the front the procedure reduces again to a single integration.

In the specifications of the case care was taken to specify the wind field close to the front. For obvious reasons of symmetry some grid points are expected to fall on the front, and it was therefore stated that wind at these points had to be considered as directed to the north. If the grid is not very narrow, this can appreciably influence the results close to the front, where most of the interest is concentrated. This problem vanishes with the ray

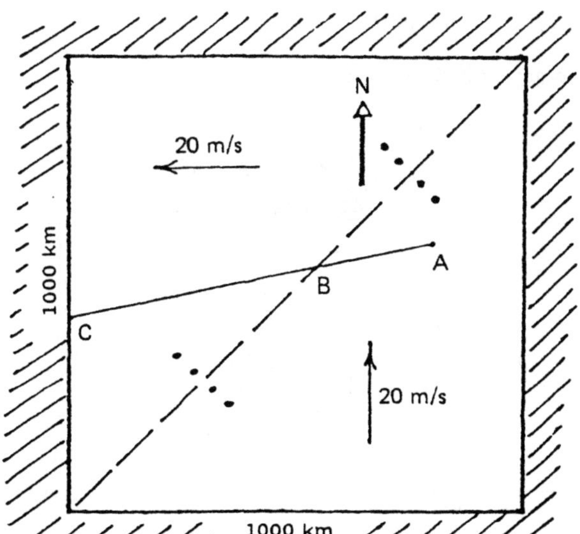

Fig. 16.5. Case V, frontal passage. The field has been covered with a net of 13×13 points. The eight dots show the points chosen for special plots.

technique, as the ray length, and each section of it, are established with continuity and the position of the front is correctly taken into account.

Figure 16.6 shows the 2d spectrum of a point just above the front. Note how energy is definitely divided into two patches. The upper one, aligned along 0°, is swell coming across the front from the area below it. The second system of waves, distributed around 90°, is aligned along the local wind direction and locally generated. Note also how, being just above the front, swell energy is still higher than that locally generated. Moving to the north away from the front the latter grows and gradually overtakes the swell in importance.

16.2.6. Case VI: Hurricane

The thin line in Fig. 16.7 shows the area where the storm has officially been defined for Case VI. The wind field is defined up to a distance of 1600 km from the storm's eye. Use of the Ross hurricane model for energy

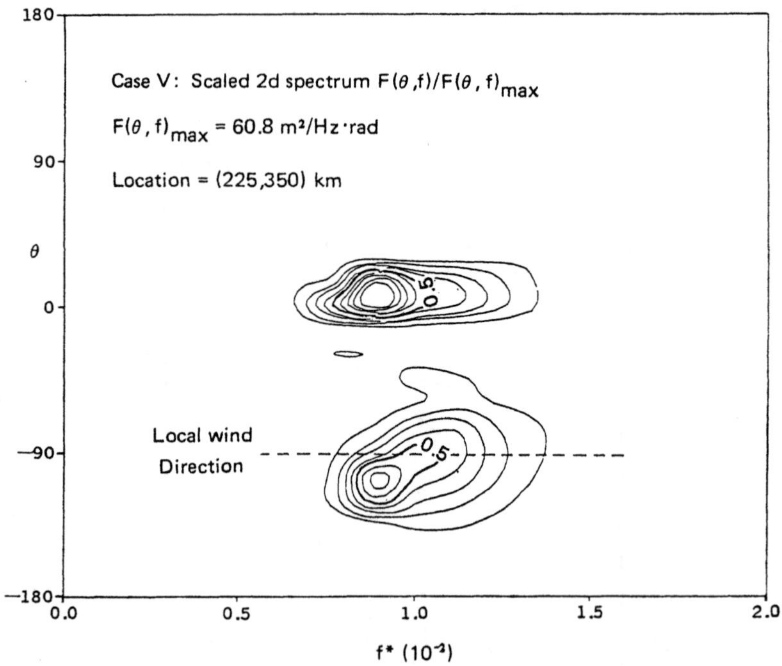

Fig. 16.6. Case V, 2d spectrum for a point just above the front. The upper energy system derives from the area below the front; the lower one, along the wind direction, is locally generated.

coming into the area from the boundaries was suggested. But this approach, because of the small grid step required by the high variation of the spatial wind field, would result in a large number of points and a significant computational effort. We have therefore chosen a different approach. Consider a target point A and a given ray (f, ϕ) leaving it. Once a system of coordinates is defined, we can extend the ray as much as necessary, as the only data to be recorded are the coordinates of the intermediate points along the ray, defined at Δt time intervals, if Δt is the time step we think suitable for the application of (16.10). The obvious approach is therefore to

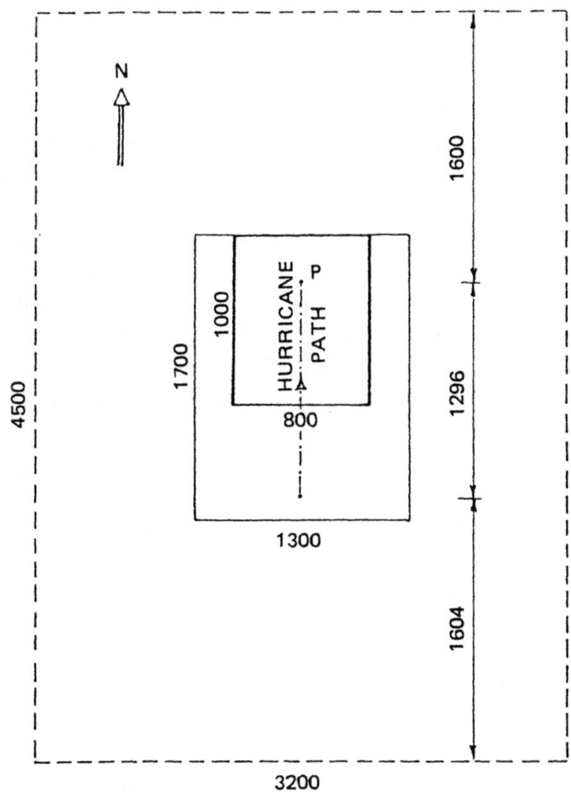

Fig. 16.7. Case VI, hurricane. The thin line shows the area originally defined for the analysis. The heavy line bounds the area from which the results were produced. For our model the area of analysis has been extended up to the broken line, where the wind is null. All the measurements are in kilometers. Grid step for results is 50 km. Special plots to be produced at 13 points around the final eye position P.

propagate the rays until out of the windy area, i.e., until more than 1600 km from any of the positions successively touched by the eye during the 24 h of the run.

This area is indicated by the broken line in Fig. 16.7, and it would obviously be difficult to cover such an area (3200×4500 km) with the 50-km-step grid suggested by the diagrams required by the test. The whole procedure is therefore summarized as follows. Once a target point is chosen, we consider all the rays, defined in frequency and direction, leaving it. Each ray is sequentially analyzed. First the coordinates of the ray points are found. This is simply done with the two expressions

$$x_i = x_0 + v \cdot \cos\phi \cdot \Delta t \cdot i$$

$$y_i = y_0 + v \cdot \sin\phi \cdot \Delta t \cdot i$$

where (x_0, y_0) are the target point coordinates and i is the index of the sequential points along the ray. The smaller Δt, of the larger the number of points and the required computer time. Therefore the value of Δt results from a tradeoff between the accuracy of the procedure and the computational effort. A suitable choice for the present case is 40 min. Direct tests showed that, while up to this value the various results from different choices of Δt varied by less than 1%, a further increase (to 1 h) would immediately produce errors of up to 5%. Once the ray (the points' coordinates) has been established, it is run backwards, from the furthest point toward the target one. At each point wind is evaluated from the given wind field and α and β coefficients are estimated accordingly. The energy at the next point is estimated by means of (16.10), and the procedure is repeated up to the final value of energy at the target position. The wind field was supplied with five different maps (each 20×20 points) with geometrically progressive resolution (from 10 to 160 km). The wind at each ray point is evaluated by choosing the smallest map including the point and using a double linear interpolation on the four grid points surrounding the actual one. For the stationary storm the wind field is centered on the final eye position, and, on evaluating the wind at each ray point, reference is made at this position. For the moving storm, time has to be taken into account. If i is the index of a point on the ray, $t' = i \cdot \Delta t$ is the time the energy patch still needs to reach the target position. Hence, if $u = 15$ m/s is the north velocity of the storm, when the energy is at the ith point, the wind field must be conceived as shifted downwards (to the south) by a distance $d = u \cdot t'$. In practice the wind field is considered as fixed in space, but, for a point with (x, y) coordinates, the actual wind is estimated as if the point were located at $(x, y + u \cdot t')$.

For the final results attention was focused on an area defined by $x = 250$–1050 km and $y = 700$–1700 km, indicated by a heavy line in Figure 16.7. This was covered with a net of 17×21 points at 50-km intervals, plus four extra points close to the final eye position P. Full calculations were carried out for all of these 365 points. Particular attention for the local plots has been devoted to the 13 points distributed around P.

The Custer diagram of Fig. 16.8 shows clearly the circular wave field (H_s) associated with a hurricane. There is a strong difference between the maximum H_s in the stationary and the moving cases (11 m and 15 m, respectively). This is mainly the result of the very large waves present in the moving hurricane and produced at the frequency running in to the north

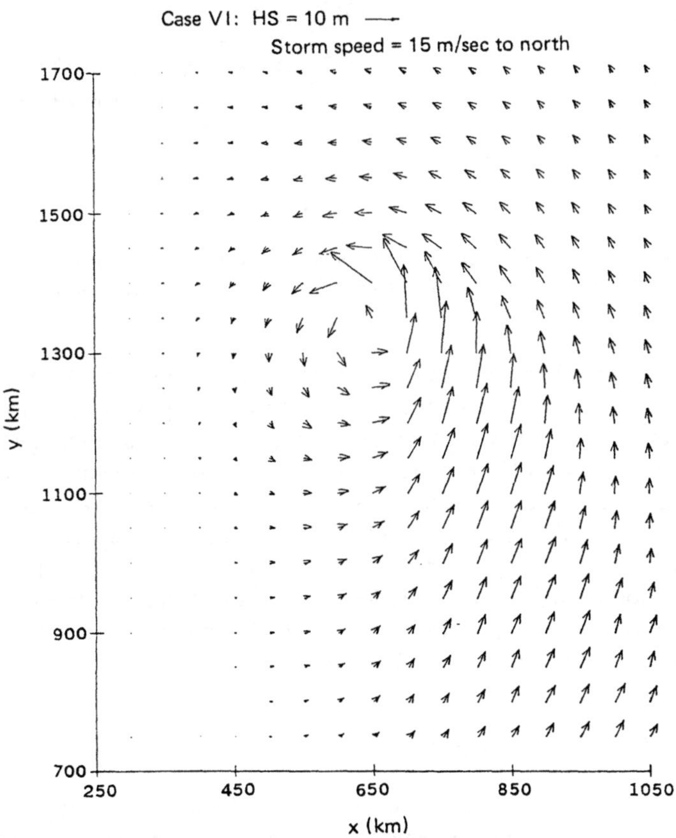

Fig. 16.8. Case VI. Custer diagram of significant wave height for the moving hurricane. Direction is evaluated by a weighted average on a 2d spectrum.

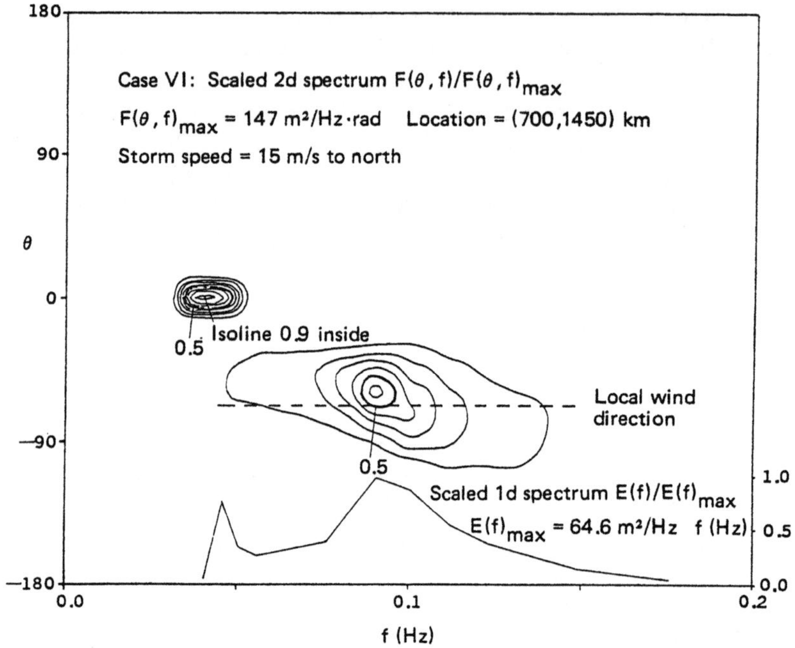

Fig. 16.9. Case VI. 2d and 1d spectra for a point close to the final position of the hurricane eye. Note the insulated component at almost 90° with respect to the local wind direction.

with a group velocity similar or close to that of the storm. An energy patch placed to the right side of the eye, and progressing with it, is permanently subjected to extremely high winds, leading to a very high energy level. The 2d spectra close to the eye are particularly complex, because of the rapidly varying wind field and the consequent presence of high wave systems running in wildly different directions. An example is reported in Fig. 16.9 together with the corresponding 1d spectrum. Note the locally generated waves, closely aligned with the local wind direction, and the very narrow, both in frequency and direction, patch of high-energy swell, running at almost 90° with respect to the local wind direction.

16.2.7. Test VII: Turning Wind

Each ray is considered split into two sections, whose length equals the group velocity v of the frequency considered times the time $T_{1,2}$; the wind is blowing to the north and to the west respectively. The final energy is estimated by a double application of (16.10), the α and β coefficients on each section being evaluated according to the respective actual wind and ray directions.

While no problem exists for the case of fully developed conditions at $t = 0$, discrete values in the integration time and the spectral frequencies considered make it difficult to evaluate precisely when the peak frequency f_p equals $2f_{PM}$, twice the Pierson–Moskowitz one. For this calculation, advantage has been taken of the adaptability of the model, choosing, as a preliminary run, a dense frequency distribution around $2f_{PM}$ and choosing for the evaluation of the spectra several times around the expected one at which $f_p = 2f_{PM}$. Time for having $f_p = 2f_{PM}$ has been estimated as $T_1 = 2.8$ h (2 h, 48 min).

For the final run the first ray section has a length $v \cdot T_1$ and the second a length $v \cdot T_2$, T_2 varying according to the specified times at which to evaluate the directional spectrum.

16.3. COMMENTS

There are a number of advantages in the ray approach to the problem of evaluating a wave spectrum. Our model is extremely flexible, and it could be adapted to the different test cases without any particular difficulty. There was a particular advantage in some of the cases owing to the uniform wind field, allowing a single- or double-step integration of the energy equation. But even in the more general case the model is highly economical. In a certain sense it is the natural solution when we want to monitor only one or a few points. But its economy is such to make it suitable for monitoring a whole area, carrying out the calculations for each point of a suitable grid in the area. This was the technique used in the study. Table 16.1 reports the computer time required by each test. The computer used was a CDC 76. Times are given as the execution time for each point and overall time, including compilation. The cases of stationary and mobile hurricanes showed no significant difference. This efficiency is of course obtained at the expense of a strong simplification of the source function. Nonlinear processes are not considered, but this is compensated for by the use of a β directly fitted to experimental data.

Besides—and we think these are the two most important points—the model was built with an eye to shallow-water applications, and it was not designed to be tuned to different experimental data from separate locations. The model has been extensively used in shallow-water studies (Cavaleri and Malanotte Rizzoli, 1981). Of course all shallow-water processes were cancelled in the present tests because of the deep-water assumption, but the model would be ready for an analogous shallow-water test as it is. As

Table 16.1
Computer Time[a]

	Case II	Case III	Case IV	Case V	Case VI	Case VII
Overall time	4.98	4.78	5.009	7.49	322.25	8.59
Execution time	0.027	0.039	0.021	0.032	0.87	0.047

[a] For each case the overall time (compilation plus execution) and the execution time for a single point are reported. The computer is a CDC 76. Times in seconds.

regards the second point, the model possesses no "tuning knob." The physics is introduced as described in the literature, and the model is run. There is certainly a good amount of empiricism in it, but we think the results support the approach.

REFERENCES

Banner, M. L., and W. K. Melville (1976): On the separation of air flow over water waves. *J. Fluid. Mech.* **77,** 825.

Barnett, T. P. (1968): On the generation, dissipation and prediction of ocean wind waves. *J. Geophys. Res.* **73,** 513.

Barnett, T. P., and J. C. Wilkerson (1967): On the generation of ocean wind waves as inferred from airborne radar measurements of fetch-limited spectra. *J. Mar. Res.* **25,** 292.

Cavaleri, L., and P. Malanotte Rizzoli (1981): Wind wave prediction in shallow water—theory and application *J. Geophys. Res.* **86,** 10961.

Collins, J. I. (1972): Prediction of shallow-water spectra. *J. Geophys. Res.* **77,** 2693.

Ewing, J. A. (1971): A numerical wave prediction model for the North Atlantic Ocean. *Dtsch. Hydrogr. Z.* **24,** 241.

Gelci, R., H. Cazalé, and J. Vassal, (1957): Prévision de la houle. La méthode des densités spectroangulaires. *Bull. Inform. Comité Central Oceanogr. d'Etude Côtes*, **9,** 416.

Hasselmann, K. (1962): On the nonlinear energy transfer in a gravity-wave spectrum. 1. General theory. *J. Fluid Mech.* **12,** 481.

Hasselmann, K. (1963): On the nonlinear energy transfer in a gravity-wave spectrum. 2. Conservation theorems, wave-particle correspondence, irreversibility. *J. Fluid Mech.* **15,** 273.

Hasselmann, K. (1974): On the spectral dissipation of ocean waves due to whitecapping. *Boundary-Layer Metorol.* **6,** 107.

Hasselmann, K., and J. I. Collins (1968): Spectral classification of finite depth gravity waves due to bottom friction. *J. Mar. Res.* **26,** 1.

Hasselmann, K., T. P. Barnett, E. Bouws, H. Carlson, D. E. Cartwright, K. Enke, J. A. Ewing, H. Gienapp, D. E. Hasselmann, P. Kruseman, A. Meerburg, P. Müller, D. J. Olbers, K. Richter, W. Sell, and H. Walden (1973): Measurements of wind-wave growth and swell decay during the Joint North Sea Wave Project (JONSWAP), *Dtsch. Hydrogr. Z., Suppl. A* **8,** (12).

Hasselmann, K., D. B. Ross, P. Müller, and W. Sell (1976): A parametric wave prediction model. *J. Phys. Oceanogr.* **6,** 200.

Miles, J. W. (1957): On the generation of surface waves by shear flows. *J. Fluid Mech.* **2,** 185.

Phillips, O. M. (1957): On the generation of waves by turbulent wind. *J. Fluid Mech.* **2,** 417.

Phillips, O. M. (1958): The equilibrium range in the spectrum of wind-generated ocean waves. *J. Fluid Mech.* **4,** 426.

Phillips, O. M. (1966): *The Dynamics of the Upper Ocean*, Cambridge University Press, London, 261 pp.

Resio, D. T. and C. L. Vincent (1977a): A numerical hindcast model for wave spectra on water bodies with irregular shoreline geometry. Part I: Test of non-dimensional growth rates. Miscellaneous Paper H-77-9, Hydraulic Laboratory, U. S. Army Engineer Waterways Experiment Station, Vicksburg, Mississippi.

Resio, D. T. and C. L. Vincent (1977b): A numerical hindcast model for wave spectra on water bodies with irregular shoreline geometry. Part II: Model verification with observed water data. Miscellaneous Paper H-77-9, Hydraulic Laboratory, U. S. Army Engineer Waterways Experiment Station, Vicksburg, Mississippi.

Willmarth, W. W., and C. E. Woolridge (1962): Measurements of the fluctuating pressure at the wall beneath a thick turbulent boundary layer. *J. Fluid Mech.* **14,** 187.

Chapter 17

The Norwegian Wave Model NOWAMO

Johannes Guddal

17.1. INTRODUCTION

The Norwegian Wave Model (NOWAMO) was developed by Odd Haug in 1966–1967 (Haug, 1968) and has been operational since then for hindcast and forecast purposes. NOWAMO is a hybrid model combining a parametric windsea development with a system of freely propagating swell components.

17.2. DEFINITIONS OF SEA AND SWELL

Sea is defined as a system of waves under continuous influence by its generating wind.

Swell is defined as a system of waves having escaped its generating wind. Swells are former windsea components, decoupled and propagating freely, but interacting with the local windsea under certain conditions.

17.3. WINDSEA CALCULATION

The windsea energy spectrum applied in the model was that proposed by Neumann and Pierson (1957). The directionality of the spectrum is described by a $\cos^2$ distribution centered on the wind direction. The one-dimensional spectrum gives the spectral density $E(f)$ in terms of the wind speed U and a minimum frequency f_{MIN}:

$$E(f) = \frac{KU}{k^4} f^{-4} \exp\left[-(Uf/k)^{-2}\right] \qquad f > f_{MIN} \qquad (17.1)$$

Johannes Guddal • The Norwegian Meteorological Institute, 5000 Bergen, Norway.

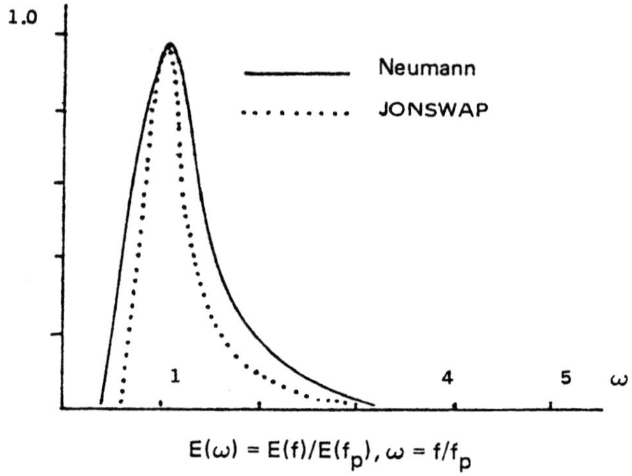

Fig. 17.1. The JONSWAP and Neumann spectra.

where $E(f) = 0$ for $f \leq f_{\mathrm{MIN}}$, $k = 4.29\ \mathrm{s^2 m^{-1}}$, and $K = 3.37\ 10^{-7}\ \mathrm{s^3 m^{-3}}$.

Comparison with the JONSWAP spectrum,

$$E(f) = \alpha g^2 (2\pi)^{-4} f^{-5} \exp\left\{-\tfrac{5}{4}(f/f_{\mathrm{m}})^{-4} + \ln\gamma \exp\left[-(f - f_{\mathrm{m}})^2/2\sigma^2 f_{\mathrm{m}}^2\right]\right\} \tag{17.2}$$

(see also Fig. 17.1) shows the Neumann spectrum to be broader and significantly less peaked.

In a growing sea state with constant wind, the cutoff frequency $f = f_{\mathrm{MIN}}$ is moved toward lower values, leaving the high-frequency tail of the spectrum unaltered. A fully developed sea is defined by $f_{\mathrm{MIN}} = 0$, or uniquely by U.

This simple spectral definition facilitates a rapid numerical integration of the energy balance equation

$$\frac{\partial E}{\partial t} + \nabla \cdot \mathbf{c}_{\mathrm{g}} E = S = S_1 + S_2 \tag{17.3}$$

where E is the total local windsea energy, $\mathbf{c}_{\mathrm{g}}$ is the mean group velocity of all spectral components, t is time, and S is the value of sources and sinks.

NOWAMO is valid for deep-water conditions only, and the source term is composed of 2 parts, S_1 and S_2:

S_1: *Air–sea energy transfer* S_1. For sea states that are not fully developed, the total sea energy is allowed to grow in the manner described above. The rate of growth is an empirical function proposed by Neumann and given some modifications by Haug (unpublished). In principle, it describes the approach of the cutoff frequency f_{MIN} toward zero as a function of U and the stage of development. In a decaying sea, the propagation part of the time-step integration may have produced a total energy E which exceeds the fully developed stage E_{MAX}, corresponding to a smaller wind speed U at the end of the time step. In this case, the excess energy is kept as a swell potential, but left out of the further integration of (17.3).

S_2: *Dissipation owing to wind shift.* According to a basic principle in the model, the directional distribution of energy is at any time uniquely determined as a function ($\cos^2$) of the angle between wind direction and partial wave propagation. After the propagation and growth/decay part of the integration step, it is generally found that a wind directional change, $\Delta\alpha$, has taken place. Applying now the $\cos^2$ directional distribution of energy density, we define the frequency-integrated energy, centered on two different wind directions, as having the difference $\Delta\alpha$ from the start to the end of the time step (Fig. 17.2). The retained energy after wind shift is taken to be defined by the overlapping area of the two distributions. This means that the preliminary computed energy E_{PREL} should be modified by a factor

$$q = 1 - \frac{|\Delta\alpha|}{\pi} - \frac{|\sin\Delta\alpha|}{\pi} \qquad |\Delta\alpha| < \pi/2$$

$$q = 0 \qquad |\Delta\alpha| \geq \pi/2$$

$$E_{RETAINED} = q \cdot E_{PREL} \tag{17.4}$$

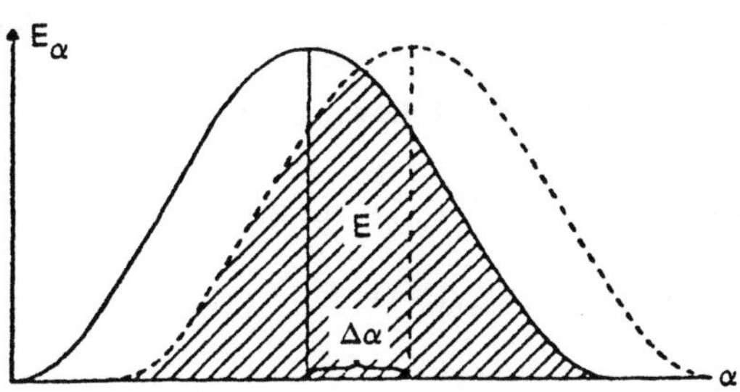

Fig. 17.2. Energy dissipation owing to wind shift.

17.4. SWELL CALCULATIONS

The swell calculation procedure is a retrospective method, applied only for selected positions. At a given time and for a selected position, all windsea energy produced within the preceding 48 h is regarded as swell potential. According to the assumed directional distribution of sea energy propagation, swell components may be localized, more or less reduced by a lack of orientation toward the actual position. Dissipation will alter the spectral shape during the propagation time 0–48 h in such a way that the high-frequency flank of the spectrum is wiped out, while the rest of the spectrum becomes narrower and the peak frequency tends to decrease. This process is governed by a frequency- and time-dependent dissipation scheme.

In this way, a large number of potential swell components may be found, all produced in the preceding 48-h period and all arriving at the same time at the actual position, after being propagated freely with their individual group velocities

$$c = \frac{g}{4\pi f} \tag{17.5}$$

A discretization process is then applied. A main swell direction is defined in the direction of the overall maximum energy component. Along this direction, final swell components are computed as the means of continuous strings of components around the main direction. This procedure utilizes the general assumption that when significant swells occur they have been generated by a well defined wind system at some distance from the actual arrival position.

17.5. COMBINED SEA AND SWELL

A simple procedure is assumed for the combination of locally generated sea and arriving swell components. The combined spectrum is computed component-by-component in the following way:

First step: $E'(f) = \max[E_s(f), E_{sw}(f)\cos^2\alpha]$

Second step: $E_{comb}(f) = E'(f) + E_{sw}(f)\sin^2\alpha$

where the subscripts s, sw, and comb mean windsea, swell, and combined sea, respectively. α is the directional difference between windsea main direction and swell main direction. A combined "main direction" is defined as either windsea or swell main direction, depending on which of these carries the highest amount of energy.

In the construction of the two-dimensional spectrum, $E(f,\theta)$, the following procedure is applied:

$$E_s' = \frac{2}{\pi} E_s(f)\cos^2(\theta - \theta_s) \qquad E_s' = 0 \quad \text{for} \quad |\theta - \theta_s| \geq \pi/2$$

$$E_{sw}' = \frac{2}{\pi} E_{sw}(f)\cos^2(\theta - \theta_{sw}) \qquad E_{sw}' = 0 \quad \text{for} \quad |\theta - \theta_{sw}| \geq \pi/2$$

where θ_s and θ_{sw} are sea and swell main directions, and θ is the arbitrary direction. Therefore

$$E(f,\theta) = \max\left[E_s', E_{sw}'\cos^2(\theta_s - \theta_{sw})\right] + E_{sw}'\sin^2(\theta_s - \theta_{sw})$$

17.6. NUMERICS

In the daily forecasting routine, NOWAMO is run in a space–time grid of resolution 150 km and 3 h. For the intercomparison studies, these values were changed to 75 km and 1.5 h. Swell components are computed only every 6 h during the 48 h period. The propagation scheme is a modification of the "Le Levier scheme" (Potter, 1973). NOWAMO applies its model spectrum defined in period bands of width 2 s, which means that an interpolation scheme is used for the conversion to frequency bands.

ACKNOWLEDGMENT. This chapter was written with the full consent of the model designer, Odd Haug.

REFERENCES

Haug, O. (1968): A numerical model for prediction of sea and swell. *Nor. Meteorol. Inst. Meteorol. Ann.* **5**(4), 139.

Neumann, G., and W. J. Pierson, Jr. (1957): A detailed comparison of theoretical wave spectra and wave forecasting methods. *Dtsch. Hydrogr. Z.* **10**, 34–146.

Potter, D. (1973): *Computational Physics*, John Wiley & Sons, London, 71.

Chapter 18

The KNMI Operational Wave Prediction Model GONO

W. J. P. de Voogt, G. J. Komen, and J. Bruinsma

18.1. INTRODUCTION

Routine wave forecasting for the southern North Sea is performed by the Royal Netherlands Meteorological Institute (KNMI) with the numerical wave model GONO (GOlven NOordzee) (Janssen *et al.*, 1984). This model was developed by Sanders (1976) in the early 1970s based on an idea by Haug (1968). Application of the model to the southern part of the North Sea necessitated the incorporation of shallow-water effects. These extensions have been recently reported by Sanders *et al.* (1980, 1981). The model has been calibrated and tested in a hindcast study of several storms (cf. Sanders *et al.*, 1980) and it has been verified for operational purposes against wave data from several positions (cf. Bouws *et al.*, 1980a, b). Furthermore, the shallow-water results in the GONO model during the winter 1979–1980 have been compared with those of the BMO model (Golding, 1978), which is operational at the British Meteorological Office (cf. Bouws *et al.*, 1980a, b, 1981).

Qualitatively, GONO is characterized by the following:

1. The model is hybrid: It uses a finite difference scheme for the propagation of the parametrical windsea and a spectral ray technique for the propagation of swell.

W. J. P. de Voogt • Delft Hydraulics Laboratory, 8300 AD Emmeloord, The Netherlands. **G. J. Komen** • Royal Netherlands Meteorological Institute (KNMI), 3730 AE De Bilt, The Netherlands. **J. Bruinsma** • Rijkswaterstaat, Deltadienst, 2597 AT The Hague, The Netherlands.

2. The total windsea energy is used as a prediction parameter.
3. The model is empirical in the sense that it uses empirical relations for the wave growth and the windsea spectral parameters.

This chapter briefly quantifies the above properties in three parts. The first part describes the growth and advection of windsea from the energy-balance equation, the second part shows how the windsea energy is fitted into a spectral shape, and the third part deals with the computation and propagation of swell. A more detailed description of GONO and its computer code is given by Bruinsma *et al.* (1980).

18.2. THE CALCULATION AND PROPAGATION OF WINDSEA

GONO starts from the version of the transport equation

$$\frac{\partial E}{\partial t} + \nabla \cdot \mathbf{c}E = B \tag{18.1}$$

where the energy $E = \int\int F(f,\theta)\, df\, d\theta$ and the energy flux $\mathbf{c}E = \int\int \mathbf{c}_g(f) F(f,\theta)\, df\, d\theta$. Here $F(f,\theta)$ is the energy density for frequency f and direction θ at x, y, and t, and $\mathbf{c}_g(f)$ is the group velocity derived from linear theory. First only windsea is considered, i.e., the part of the spectrum which grows actively under the influence of the wind or is just in equilibrium with the wind. For pure windsea a universal angular distribution is assumed:

$$F_{\mathrm{ws}}(f,\theta) = \frac{2}{\pi} F_{\mathrm{ws}}(f) \cos^2(\theta - \phi) \qquad |\theta - \phi| \le \frac{\pi}{2} \tag{18.2}$$

where ϕ is the local wind direction. The source term B incorporates the net atmospheric input and, in shallow water, energy dissipation.

For every time step $\Delta t = 1\frac{1}{2}$ h, equation (18.1) is solved in two parts. The advective part, $\nabla \cdot \mathbf{c}E$, is integrated in a Cartesian grid with spacings $\Delta x = \Delta y = 75$ km by means of a first-order finite-difference scheme (time forward, upwind). This scheme is energy-conserving, but shows under conditions of constant propagation speed an energy spread S given by

$$S^2 = n(\Delta y)^2(aC - bC^2) \tag{18.3}$$

where n is the number of time steps, C is the Courant number, and the constants a and b are equal to 1.207 and $\simeq 0.721$, respectively.

After each advection step the average direction of propagation of the new energy in each grid point is calculated. Then the new energy is reduced by a factor which takes a value between zero and unity depending on the difference between the average propagation direction and the local wind direction. The reduced energy is taken as the new windsea in the local wind direction. The "lost" energy is still present in the swell scanned from the windsea at the previous timestep. The local-growth contribution to the solution of equation (18.1), B, is found for deep water from a time-limited empirical growth curve which is shown in nondimensional form in Fig. 18.1. Details of the solution for depth-limited growth can be found in Sanders *et al.* (1981). The deep-water growth curve is also used to define the stage-of-development parameter

$$\xi = (E/E_{\max})^{1/4} \tag{18.4}$$

where E is the actual windsea energy and $E_{\max}$ its maximum value for a given windspeed, u. The importance of this parameter will be explained in the next section.

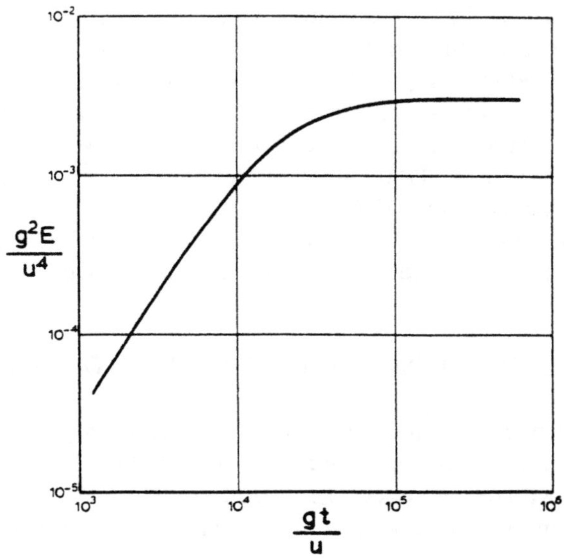

Fig. 18.1. The empirical growth curve of GONO as a function of time for infinite fetch and deep water.

18.3. THE ENERGY DISTRIBUTION FOR WINDSEA

The one-dimensional windsea spectrum in GONO takes the parameterized form (Kruseman, 1976)

$$F_{ws}(f) = 0 \qquad 0 \leq f \leq f_{min}$$

$$F_{ws}(f) = \frac{\hat{\alpha} g^2}{(2\pi)^4} \frac{1}{f_p^5} \frac{f - f_{min}}{f_p - f_{min}} \qquad f_{min} < f \leq f_p \tag{18.5}$$

$$F_{ws}(f) = \frac{\hat{\alpha} g^2}{(2\pi)^4} \frac{1}{f^5} \qquad f > f_p$$

where $\hat{\alpha}$ is the Phillips "constant," f_p is the frequency of peak energy density, and f_{min} is the minimum frequency. These parameters are coupled via the total energy, viz.,

$$E_{ws} = \int_{f_{min}}^{\infty} F_{ws}(f)\, df = \frac{\hat{\alpha} g^2}{(2\pi)^4} \frac{1}{4 f_p^4}(3 - 2\mu) \qquad \mu = f_{min}/f_p \tag{18.6}$$

The energy of the sea state in a grid point is fitted into the above spectral shape according to the following procedure. First, ξ [equation (18.4)] is obtained from E_{ws} and the local wind speed, u. Then $\hat{\alpha}$ and f_p are calculated from power-law relations between $\hat{\alpha}$ and ξ, and between $\hat{\alpha}$ and uf_p/g. These relations are obtained from fits to KNMI and JONSWAP(A) data (cf. Sanders *et al.*, 1980, 1981; Müller, 1976). Finally, the parameter f_{min} follows from equation (18.6).

The same procedure is followed for shallow water, but there the stage of development is corrected for bottom influence (cf. Sanders *et al.*, 1981).

18.4. THE CALCULATION AND PROPAGATION OF SWELL

GONO calculates swell for a number of prechosen grid points, called swell points. The calculation is done for six directional sectors and seven period bands with center periods T_m ranging from 7 to 19 s. To that end, for each swell point **X**, circles of radii R and center **X** are defined for each T_m by

$$R(T_m) = c_g(T_m) n\, \Delta t \qquad 1 \leq n \leq 47 \tag{18.7}$$

where $\mathbf{c}_g(T)$ is the group velocity for deep water and $\Delta t = 1\frac{1}{2}$ h. Thus, in deep water, swell trains of period T_m located at the circle and propagating toward $\mathbf{X}$ arrive at $\mathbf{X}$ after $n\,\Delta t$ h. GONO propagates these swell trains along rays i of length $|\mathbf{R}_i| = R(T_m)$ and direction θ_i $(i = 1, \ldots, N)$. In order to have the optimal accuracy for swell of both near and distant origin the total number of rays, N, depends linearly on $R(T_m)$. This dependency is chosen such that the distance between the origins on the circle of adjacent rays is approximately equal to the spacing of the Cartesian grid. The above definitions are illustrated in Fig. 18.2 for a particular value of n.

The basic idea of GONO is to compute the directional wave spectrum in swell point $\mathbf{X}$ at time t as a sum of swell and local windsea. To see if a directional sector contains swell energy the concept of "potential swell" is introduced. Potential swell is the amount of wave energy in the period band around T_m in directional sector j that would reach $\mathbf{X}$ at time t if the wind would have dropped at time $t - \tau$:

$$E_j^{\text{pot}}(\mathbf{X}, t;\ \tau) \tag{18.8}$$

Suppose E_j^{pot} is known for $\tau_{n+1} = (n+1)\,\Delta t$, and that it is wanted for $\tau_n = n\,\Delta t$. Then two possibilities have to be considered: Either the wave components that will arrive at $\mathbf{X}$ and time t do not grow between $t - \tau_{n+1}$ and $t - \tau_n$, or they do. In the first case E_j^{pot} remains unchanged (apart from

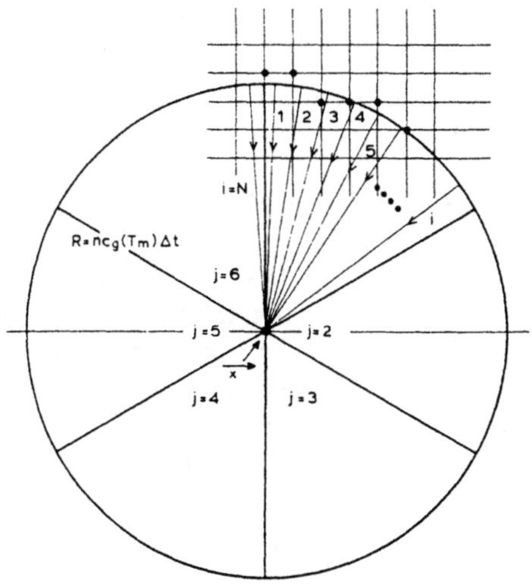

Fig. 18.2. The swell scanning procedure of GONO.

possible dissipation). In the second case E_j^{pot} increases, but then it can be calculated from the sea state at time $t - \tau_n$. So,

$$E_j^{\text{pot}}(\mathbf{X}, t;\ \tau_n) = \max\left[E_j^{\text{ws}}(\mathbf{X}, t;\ \tau_n), E_j^{\text{pot}}(\mathbf{X}, t;\ \tau_{n+1})\right] \tag{18.9}$$

where E_j^{ws} is the wave energy of period T_m in sector j that would reach $\mathbf{X}$ at time t, as calculated from the *windsea* state at $t - \tau_n$ with *no* winds for times larger than $t - \tau_n$. Thus,

$$E_j^{\text{ws}}(\mathbf{X}, t;\ \tau_n) = \int\int F_{\text{ws}}\left(f, \theta;\ \mathbf{X} - \mathbf{c}_{\text{g}}(f, \theta) n\,\Delta t, t - n\,\Delta t\right) df d\theta \tag{18.10}$$

where the integral is over one period band and one directional sector.

The integral is estimated in GONO by

$$\sum E_i = \sum \int_{1/(T_m+1)}^{1/(T_m-1)} \int_{\theta_i - \frac{\pi}{N}}^{\theta_i + \frac{\pi}{N}} F_{\text{ws}}(f, \theta)\, df d\theta \tag{18.11}$$

where F_{ws} is given by equations (18.2) and (18.5) and is taken at the grid point nearest to the starting point of ray i. The summation is made over all rays i in sector j. The mean direction $\bar{\theta}$ of E_j^{ws} is subsequently defined by

$$\bar{\theta} = \arctan\left(\frac{\sum E_i \sin\theta_i}{\sum E_i \cos\theta_i}\right) \tag{18.12}$$

By induction one obtains from equation (18.9) the potential swell in sector j at $t - \Delta t$ according to

$$E_j^{\text{pot}}(\mathbf{X}, t;\ \tau_1) = \max\left[E_j^{\text{ws}}(\mathbf{X}, t;\ \tau_n)\right]_{n=1}^{n=\infty} \tag{18.13}$$

Here use has been made of the fact that $E_j^{\text{pot}}(\mathbf{X}, t;\ \tau_{n+1})$ can be taken equal to zero for sufficiently large n owing to the finiteness of the ocean. The average direction of $E_j^{\text{pot}}(\mathbf{X}, t;\ \tau_1)$ is taken as the mean direction of the largest $E_j^{\text{ws}}(\mathbf{X}, t;\ \tau_n)$. Hindcast studies proved it necessary to attenuate $E_j^{\text{pot}}(\mathbf{X}, t;\ \tau_1)$ by an empirical factor in such a way that short-period energy from distant sources is attenuated more strongly than long-period energy from near sources. The total wave energy in the period band around T_m in directional sector j at location $\mathbf{X}$ and time t is finally given by

$$E_j(\mathbf{X}, t) = \max\left[E_j^{\text{ws}}(\mathbf{X}, t), E_j^{\text{pot}}(\mathbf{X}, t;\ \tau_1)\right] \qquad \text{if } |\bar{\theta} - \phi| \leq \frac{\pi}{6} \tag{18.14a}$$

and

$$E_j(\mathbf{X}, t) = E_j^{\text{ws}}(\mathbf{X}, t) + E_j^{\text{pot}}(\mathbf{X}, t;\ \tau_1) \qquad \text{otherwise} \qquad (18.14\text{b})$$

Thus, swell trains more than 30° off the local wind direction ϕ keep their identity. This provides an effective directional relaxation in turning winds. For swell within 30° of the local wind direction the problem of double counting is taken into consideration. If $E_j^{\text{ws}} > E_j^{\text{pot}}$ it is assumed that the potential swell was already contained in the local windsea, and if $E_j^{\text{ws}} \leq E_j^{\text{pot}}$ the ray calculation is supposed to be more accurate than the windsea propagation scheme.

The same procedure is followed for the calculation of swell in water of variable depths, but then the group velocity in equation (18.7) and (18.10) is replaced by the group velocity in shallow water. Consequently, the circles of equation (18.7) are transformed into closed curves $\mathbf{R}(T_m)$ which connect the starting points of the rays. These curves are defined by

$$\int_{\mathbf{X}}^{\mathbf{X}+\mathbf{R}(T_m)} \frac{ds}{c_g(T_m, D(s))} = n\,\Delta t \qquad (18.15)$$

where $D(s)$ is the depth varying along the straight path s. The lengths of the rays $|\mathbf{R}_i|$ are thus determined by the condition of equal propagation time for the energy. The number of rays is as in deep water. The swell trains are furthermore attenuated by a linear bottom dissipation rate integrated along $\mathbf{R}_i$.

The above method of swell calculation gives the correct result for distant localized source regions and has also proven to be useful in more general situations with nonuniform and nonstationary wind fields.

ACKNOWLEDGMENTS. The authors wish to acknowledge useful discussions with Peter Janssen and Jan Sanders. They thank Mr. H. de Froe for his assistance in running the computer programs for the intercomparison studies. This work is part of a joint wave modeling program of KNMI, Rijkswaterstaat, and Delft Hydraulics Laboratory.

REFERENCES

Bouws, E., B. W. Golding, G. J. Komen, H. H. Peeck, and M. J. M. Saraber (1980a): Preliminary results on a comparison of shallow-water wave predictions. Scientific Report 80-5, Royal Netherlands Meteorological Institute, De Bilt, The Netherlands.

Bouws, E., G. J. Komen, R. A. Van Moerkerken, H. H. Peeck, and M. J. M. Saraber (1980b): A comparison of shallow-water wave predictions. Scientific Report V-362, Royal Netherlands Meteorological Institutes, De Bilt, The Netherlands.

Bouws, E., G. J. Komen, R. A. Van Moerkerken, H. H. Peeck, and M. J. M. Saraber (1981): An evaluation of operational wave forecasts on shallow water. *Proc. Symp. Wave Dynamics and Radio Probing of the Ocean Surface*, Miami, 1981.

Bruinsma, J., P. A. E. M. Janssen, G. J. Komen, H. H. Peeck, M. J. M. Saraber, and W. J. P. de Voogt (1980): Description of the KNMI operational wave forecast model GONO. Scientific Report 80-8, Royal Netherlands Meteorological Institutes, De Bilt, The Netherlands.

Golding, B. W. (1978): A depth-dependent wave model for operational forecasting. *Turbulent Fluxes through the Sea Surface, Wave Dynamics, and Prediction* (A. Favre and K. Hasselmann, eds.), Plenum Press, New York, 593–606.

Haug, O. (1968): A numerical model for prediction of sea and swell. *Nor. Meteorol. Inst. Meteorol. Ann.* **5**(4), 139–161.

Janssen, P. A. E. M., G. J. Komen, and W. J. P. de Voogt (1984): An operational coupled hybrid wave prediction model. *J. Geophys. Res.* **89**, 3635–3654.

Kruseman, P. (1976): Twee practische methoden voor het maken van verwachtingen van golfcomponenten met perioden tussen 10 en 25 seconden nabij Hoek van Holland. Scientific Report WR 76-1, Royal Netherlands Meteorological Institute, De Bilt, The Netherlands.

Müller, P. (1976): Parameterization of one-dimensional wind wave spectra and their dependence on the state of development. *Hamb. Geophys. Einzelschr.* **31**.

Sanders, J. W. (1976): A growth-stage scaling model for the wind-driven sea. *Deutsch. Hydrogr. Z.* **29**, 136–161

Sanders, J. W., W. J. P. de Voogt, and J. Bruinsma (1980): Fysisch golfonderzoek Noordzee. Scientific Report MLTP-2, Royal Netherlands Meteorological Institute, De Bilt, The Netherlands.

Sanders, J. W., and J. Bruinsma (1981): Modeling wind-driven sea in shallow water. *Proc. Symp. Wave Dynamics and Radio Probing of the Ocean Surface*, Miami, Florida, 1981.

Chapter 19

The TOHOKU Wave Model

Yoshiaki Toba, Sanshiro Kawai,† and Paimpillil S. Joseph

The TOHOKU model is a hybrid model, since it combines a single-parameter growth equation for wind wave with swell components and includes an interchange between swells and wind waves.

19.1. BACKGROUND OF THE SINGLE-PARAMETER WIND WAVE PART

The basic characteristic of the wind-wave part is the use of a self-similarity structure in growing wind waves, expressed by a simple power law between the dimensionless wave height and the period for growing wind waves proposed by Toba (1972):

$$H_s^* = BT_s^{*\,3/2} \qquad B = 0.062 \tag{19.1}$$

where $H_s^* \equiv gH_s/u_*^2$ and $T_s^* \equiv gT_s/u_*$ are nondimensional forms of the significant wave height H_s and the significant wave period T_s, normalized by the acceleration of gravity g and the friction velocity of the air u_*. This power law is considered to have been substantiated by many data sets both in the laboratory and in the field (e.g., Kawai *et al.*, 1977; Mitsuyasu *et al.*, 1980). Not only is the law consistent with empirical formulas for short

Yoshiaki Toba, Sanshiro Kawai,† and Paimpillil S. Joseph • Geophysical Institute, Tohoku University, Physical Oceanography Laboratory, Sendai 980, Japan. *Present address for P.S.J.*: Center for Earth Science Studies, Regional Center, Kerala State, 682018 India. Sanshiro Kawai died on November 21, 1981.

fetches given by Wilson (1965) and by Mutsuyasu *et al.* (1971), but also Wilson's (1965) complicated formulas for long fetches give substantially the same power law on eliminating the nondimensional fetch. A comparison of (19.1) with a composite data set is shown in Fig. 19.1, taken from Kawai *et al.* 1977).

The 3/2 power law (19.1) is transformed to

$$E^* = B_f f_p^{*\,-3} \qquad B_f = 2.1\times10^{-4} \tag{19.2}$$

where $E^* \equiv g^2E/u_*^4$, $f_p^* \equiv u_* f_p/g$, and f_p the peak frequency is

$$f_p = (1.05T_s)^{-1} \tag{19.3}$$

and

$$E = \int_0^\infty \phi(f)\,df = H_s^2/16 \tag{19.4}$$

were used with $\phi(f)$ as the spectral density (Toba, 1978).

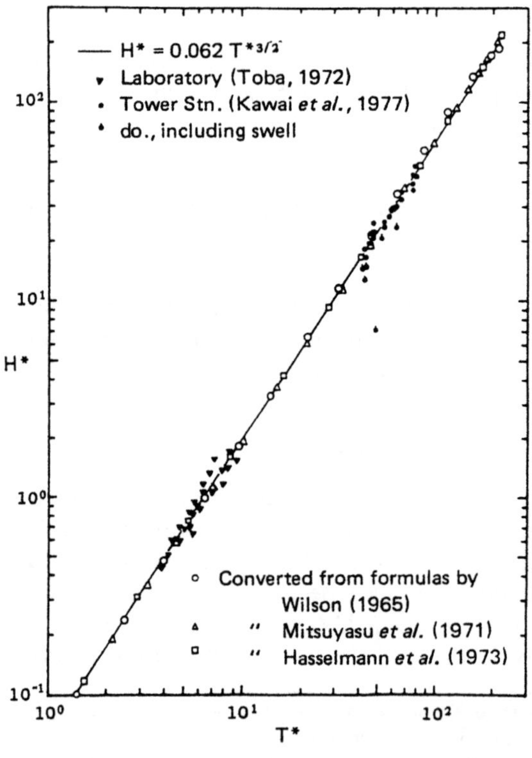

Fig. 19.1. The 3/2 power law (19.1) between nondimensional significant wave height and significant period with a composite data set. (From Kawai *et al.*, 1977.)

Equation (19.2) may be rewritten as

$$\varepsilon = 7.1 \times 10^{-6}\, \nu^{-3} \tag{19.5}$$

in the same notation as the JONSWAP formula (Hasselmann *et al.*, 1976):

$$\varepsilon = 5.3 \times 10^{-6}\, \nu^{-10/3} \tag{19.6}$$

where $\varepsilon \equiv g^2 E / U_{10}^4$ and $\nu \equiv U_{10} f_p / g$. The two formulas (19.5) and (19.6) are not significantly different, though our formula (19.5) fits slightly better a composite data set presented by Hasselmann *et al.* (Toba, 1978).

By use of the power law (19.2) along with the Wilson's fetch graph formula, a growth equation for wind waves in terms of the nondimensional total energy E^*, with a source function having a form of a simple stochastic error function, was expressed as

$$\frac{\partial(E^{*2/3})}{\partial t^*} + \frac{E^{*1/3}}{a}\frac{\partial(E^{*2/3})}{\partial x^*} = G_0 R\left[1 - \mathrm{erf}(bE^{*1/3})\right] \tag{19.7}$$

as described in detail in Toba (1978). Here

$$\mathrm{erf}(\zeta) = (2/\sqrt{\pi})\int_0^{\zeta} \exp(-\xi^2)\, d\xi$$

and $a = 0.74$, $G_0 R = 2.4 \times 10^{-4}$ and $b = 0.12$. An integration of (19.7) is shown in Fig. 19.2, cited from Toba (1978). In the present model, its total-derivative form (Kawai *et al.*, 1979),

$$\frac{d(E^{*2/3})}{dt^*} = G_0 R\left[1 - \mathrm{erf}(bE^{*1/3})\right] \tag{19.8}$$

is used. It should be noted that in (19.7) or (19.8),

$$G \equiv G_0\left[1 - \mathrm{erf}(bE^{*1/3})\right] \qquad G_0 = 0.062 \tag{19.9}$$

represents the momentum retention factor. This G_0 of 6% is the value for very young waves.

As to the spectral form of the wind wave part, our model assumes a complete self-similarity form. In the one-dimensional form, the high-frequency side of the spectrum is expressed, according to Toba (1973, 1978), as

$$\phi(f) = (2\pi)^{-3} \alpha_s g_* u_* f^{-4} \qquad \text{for } f \geq f_p \tag{19.10}$$

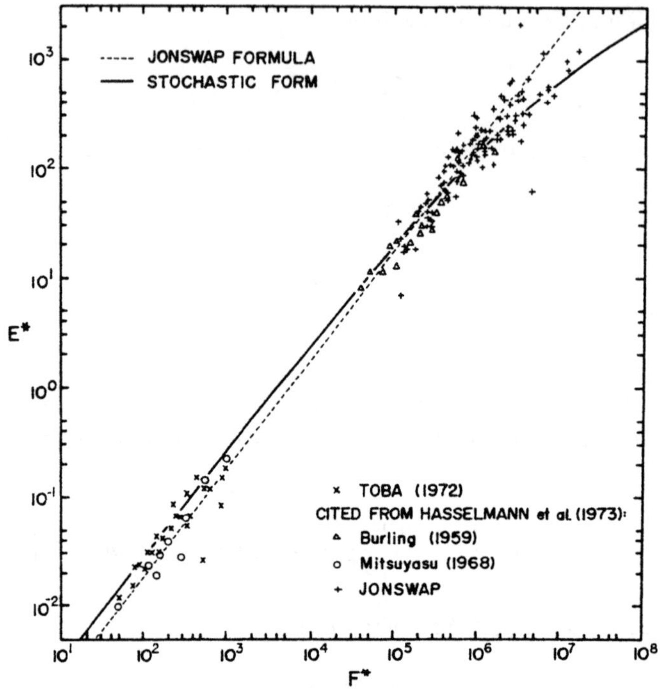

Fig. 19.2. An integration of (19.7) with some data. The E^* is nondimensional energy, and F^* is nondimensional fetch. (From Toba, 1978.)

where $g_* \equiv g(1 + Sk^2/\rho_w g)$ is g extended to include the surface tension S for the gravity-capillary range, k is the wavenumber, and ρ_w is the density of water.

Many wave models use a $g^2 f^{-5}$-type spectral form:

$$\phi(f) = (2\pi)^{-4} \alpha g^2 f^{-5} \qquad \text{for } f \geq f_p \tag{19.11}$$

which originates from Phillips (1958). However, Hasselmann *et al.* (1976) showed empirically that

$$\lambda \equiv \varepsilon \nu^4/\alpha \equiv E f_p^4/\alpha g^2 = 1.6 \times 10^{-4} \tag{19.12}$$

was a constant. Using (19.6), they proposed that

$$\alpha = 0.032 \nu^{2/3} \tag{19.13}$$

But substituting (19.5) into (19.12), it follows that

$$\alpha = 0.044\nu \tag{19.14}$$

Again there is no significant difference between the two forms, though (19.14) fits slightly better a composite data set presented by Hasselmann *et al.* (Toba, 1978). Substituting (19.14) into (19.11), since α is proportional to $U_{10}f_p/g$ in (19.14), it follows that the ϕ value for the peak frequency is proportional to $gU_{10}f_p^{-4}$. If we assume spectral self-similarity and the proportionality of U_{10} and u_*, the high-frequency side of $\phi(f)$ should be proportional to gu_*f^{-4}.

Thus the value of α_s in our spectral form (19.10) becomes very simply a constant, instead of a function of the fetch as α in (19.13). We can determine the value of α_s by assuming the self-similarity of our symmetrical spectral form on a logarithmic diagram (cf. Joseph *et al.*, 1981b) as

$$\alpha_s = 0.096 \tag{19.15}$$

Kawai *et al.* (1977) proposed a value of $\alpha_s = 0.062 \pm 0.010$, and Mitsuyasu *et al.* (1980) proposed a value of α_s ranging from 0.06 to 0.12 as a slightly varying function of the nondimensional fetch. The above value of α_s just corresponds to the midvalue of the range proposed by Mitsuyasu *et al.*

Mitsuyasu *et al.* (1980) reported quasi-equivalence of the JONSWAP spectrum and Toba's spectrum at the high-frequency side. Another aspect of quasi-equivalence was already seen between (19.5) and (19.6), and our model thus uses the simplest form among empirically relevant expressions.

Though the prediction calculation for the wind wave part is made one-dimensionally, a local two-dimensional spectrum is assumed using a simple angular distribution function of $\cos^2(\theta - \bar{\theta})$, where $\bar{\theta}$ is the direction of main energy propagation.

19.2. CHARACTERISTIC POINTS IN NUMERICAL SCHEME OF THE MODEL, INCLUDING SWELLS

The set of basic equations of the wind wave part is thus (19.8), (19.4), (19.1), and (19.3). The main flow diagram of the numerical scheme is shown in Fig. 19.3. We can explain this diagram from the integration of (19.8) for the wind wave part. The wave energy advances along the wind direction during the interval Δt by a distance $\Delta x = g(T_i + T_{i+1})\,\Delta t/8\pi$ determined by its group velocity. Then an interpolation of the energy to the grid point is performed (cf. Joseph *et al.*, 1981a).

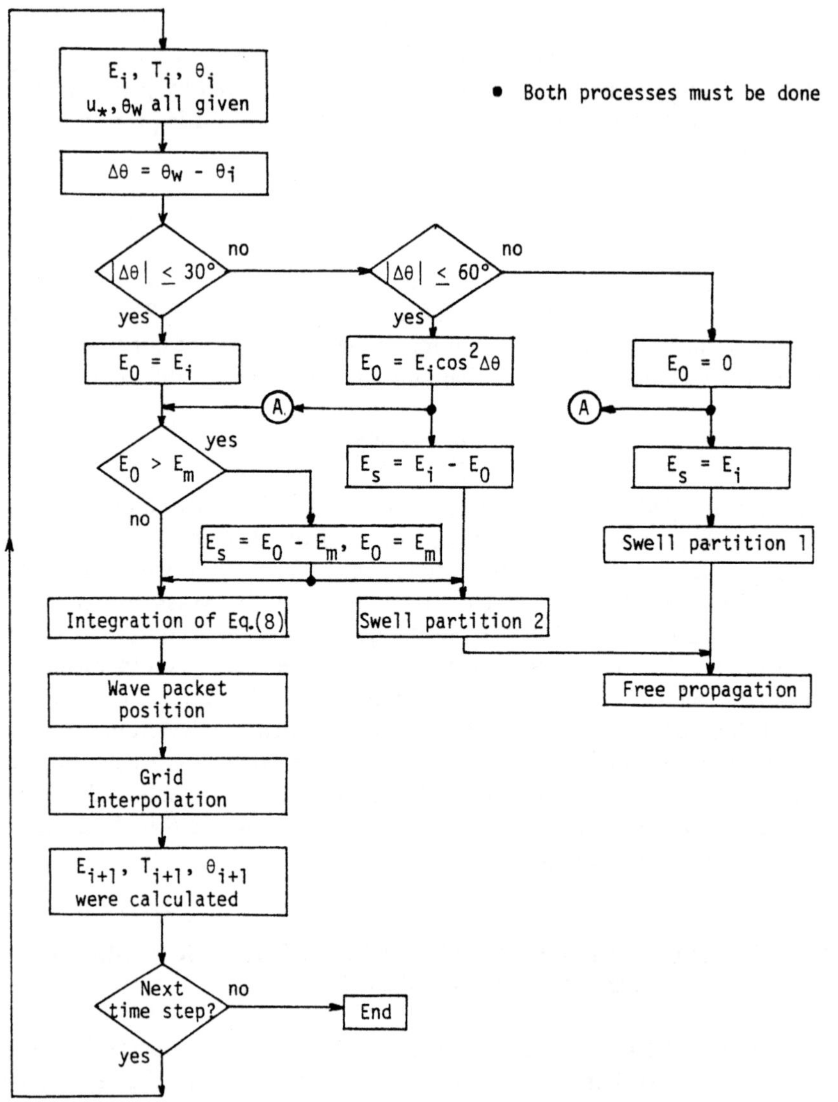

Fig. 19.3. The main flow diagram for the TOHOKU model.

The initial values are determined for the next time step to advance the process in the following way. If the wind is varying, the adjustment of wave energy is performed for three ranges of $|\Delta\theta| = \theta_w - \theta_i$, the angular deviation of the wind from the present wave propagation direction. For $|\Delta\theta|$ not more than 30°, all the wave energy is assumed to propagate as a wind wave in the new wind direction. For the range of $30° < |\Delta\theta| \leq 60°$, the energy of the wind wave part, $E_0 = E_i \cos^2 \Delta\theta$, propagates in the new direction, and the remaining energy $E_s = E_i - E_0$ moves as swells in the old direction (swell partition 2, as shown later). For $|\Delta\theta| > 60°$, the entire wave energy is considered to become swell $E_s = E_i$ in the old direction (swell partition 1).

The wind wave part is then tested against a criterion for a fully developed sea of $E_0 > E_m$ of wind waves, where E_m corresponds to $E^* = 3.7 \times 10^3$, also to $C/U_{10} = 1.37$ or $T^* = 248$. If the wind is strong, so that $E_0 \leq E_m$, the next integration of the growth equation is performed, but if the wind is weak, with $E_0 > E_m$, the excess energy $E_s = E_0 - E_m$ is transformed to the swells (also swell partition 2), and the E_m remains as the energy of wind waves.

Swells are assumed to have one-directional nature with the selected frequency components ranging from 0.04 to 0.15 Hz in 0.01-Hz frequency bands, as shown in Fig. 19.4. The energy density at the formation time

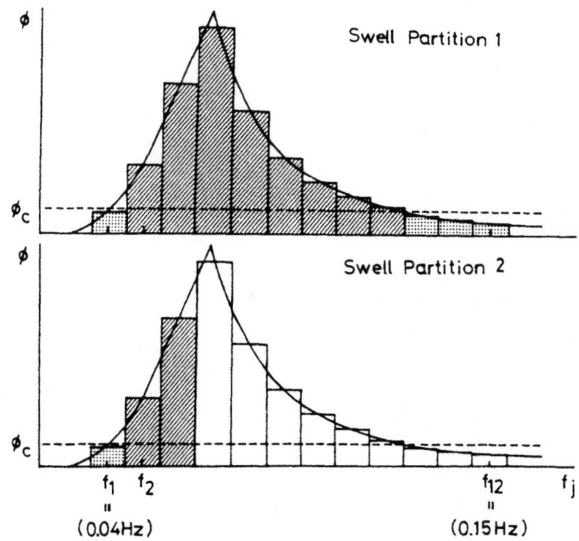

Fig. 19.4. A schematic illustration of swell partitions 1 and 2.

corresponds to the one-dimensional spectrum of wind waves. For $f > 0.15$ Hz, the energy is cut off since the value is small. In swell partition 1, only the shaded components in Fig. 19.4 are considered. The ϕ_c is an energy cutoff level included in order to neglect minor swell packets. A value of $\phi_c = 6.25\ \mathrm{m^2 Hz^{-1}}$ is adopted, corresponding to such a condition that, if the energy is confined to a single frequency band, H_s does not exceed 1 m. In swell partition 2, only the shaded energy nearly equal to E_s is propagated as swells, and the white area remains as energy of wind waves. The dotted portion remains below the cutoff energy level.

The scheme for the free propagation of swells is shown in a flow diagram in Fig. 19.5. The subscript j is the frequency number. After the phase speed C is estimated, readjustment is made by use of the wind at the nearest grid. Namely, for an adverse wind of $|\Delta\theta| > 90°$, swells propagate in their original direction but decaying by Inoue's (1967) formula, and for other $\Delta\theta$ values, as indicated.

In the output stage for the swell packets, 12 direction bins of 30° are used, and for a particular bin the swell component having a maximum energy in each frequency band is adopted and is assumed to take the midangle of the bin.

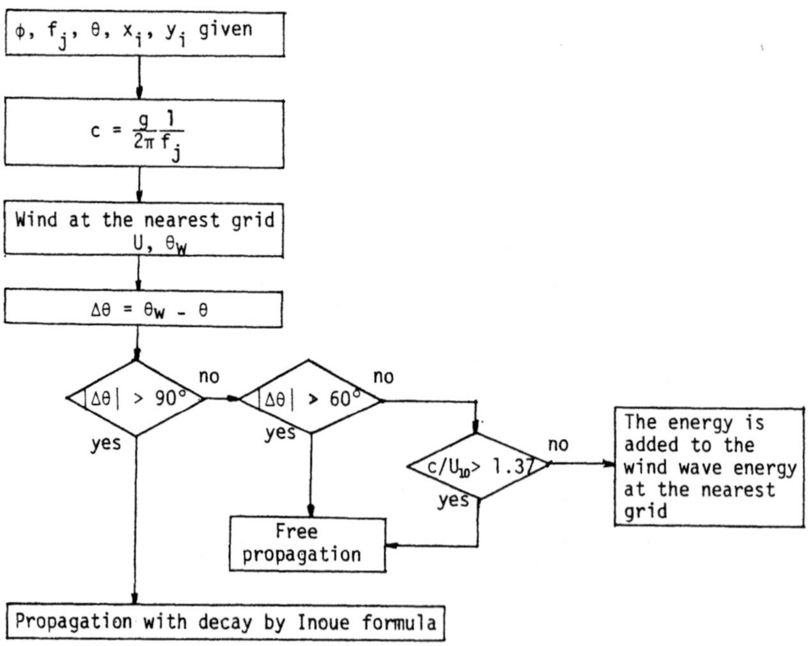

Fig. 19.5. A scheme for free propagation of swells, including reentry of energy to wind waves.

It should be noted that our model uses nondimensional parameters normalized by u_* instead of U_{10}. Consequently, prediction from a given field of U_{10} requires the form of a drag coefficient c_D defined by $U_* = (c_D)^{1/2}U_{10}$. In our model we have used c_D values in a combination of formulas by Wu (1969) and Wu (1980):

$$c_D = (0.8 + 0.065U_{10}) \times 10^{-3} \qquad \text{for } U_{10} < 27.7 \text{ ms}^{-1} \qquad (19.16)$$

and

$$c_D = 2.6 \times 10^{-3} \qquad \text{for } U_{10} \geq 27.7 \text{ ms}^{-1} \qquad (19.17)$$

More detailed descriptions of the scheme with some applications may be found in Joseph *et al.* (1981b).

19.3. CONCLUDING REMARKS

Since the wind wave part uses the simplest equations among the empirically relevant expressions, our model is the most economical one. However, the present numerical scheme, including the interchange between wind waves and swells, has some shortcomings. The main points are first that the wind wave part does not allow energy spreading aside from the wind direction, and second that energy conservation is violated, especially near a sharp front. Some modification of the scheme is now being undertaken relating to these points.

ACKNOWLEDGMENT. This work was partially supported by a Grant-in-Aid for Scientific Research from the Ministry of Education, Science and Culture, Project No. 502505.

REFERENCES

Hasselmann, K., D. B. Ross, P. Müller, and W. Sell (1976): A parametric wave prediction model. *J. Phys. Oceanogr.* **6**, 200–228.

Hasselmann, K., T. P. Barnett, E. Bouws, H. Carlson, D. E. Cartwright, K. Enke, J. A. Ewing, H. Gienapp, D. E. Hasselmann, P. Kruseman, A. Meerburg, P. Müller, D. J. Olbers, K. Richter, W. Sell, and H. Walden (1973): Measurements of wind-wave growth and swell decay during the Joint North Sea Wave Project (JONSWAP). *Dtsch. Hydrogr. Z., Suppl. A.* **8**(12).

Inoue, T. (1967): On the growth of the spectrum of a wind generated sea according to a modified Miles–Phillips mechanism and its application to wave forecasting. Report TR-67-5, New York University Geophysical Sciences Laboratory, 74 pp.

Joseph, P. S., S. Kawai, and Y. Toba (1981a): Prediction of ocean waves based on the single-parameter growth equation of wind waves. II. Introduction of grid method. *J. Oceanogr. Soc. Jpn.* **37**, 20–31.

Joseph, P. S., S. Kawai, and Y. Toba (1981b): Ocean wave prediction by a hybrid model—Combination of single-parameterized wind waves with spectrally treated swells. *Sci. Rep. Tohoku Univ. Ser. 5 (Tohoku Geophys. J.)* **28**, 27–45.

Kawai, S., K. Okada, and Y. Toba (1977) Field data support of three-seconds power law and $gu_* \sigma^{-4}$-spectral form for growing wind waves. *J. Oceanogr. Soc. Jpn.* **33**(3), 137–150.

Kawai, S., P. S. Joseph, and Y. Toba (1979): Prediction of ocean waves based on the single-parameter growth equation of wind waves. *J. Oceanogr. Soc. Jpn.* **35**, 151–167.

Mitsuyasu, H., R. Nakayama, and T. Komori (1971): Observations of the wind and waves in Hakata Bay. *Rep. Res. Inst. Appl. Mech., Kyushu Univ.* **19**, 37–74.

Mitsuyasu, H., F. Tasai, T. Suhara, S. Mizuno, M. Ohkusu, T. Honda, and K. Rikiishi (1980): Observation of the power spectrum of ocean waves using a cloverleaf buoy. *J. Phys. Oceanogr.* **10**, 286–296.

Phillips, O. M. (1958): The equilibrium range in the spectrum of wind-generated ocean waves. *J. Fluid Mech.* **4**, 426–434.

Toba, Y. (1972): Local balance in the air–sea boundary processes. I. On the growth process of wind waves. *J. Oceanogr. Soc. Jpn.* **28**, 109–120.

Toba, Y. (1973): Local balance in the air–sea boundary processes. III. On the spectrum of wind waves. *J. Oceanogr. Soc. Jpn.* **29**, 209–220.

Toba, Y. (1978): Stochastic form of the growth of wind waves in a single-parameter representation with physical implications. *J. Phys. Oceanogr.* **8**, 494–507.

Wilson, B. W. (1965): Numerical prediction of ocean waves in the North Atlantic for December 1959. *Dtsch. Hydrogr. Z.* **18**, 114–130.

Wu, J. (1969): Wind stress and surface roughness at air–sea interface. *J. Geophys. Res.* **74**, 444–455.

Wu, J. (1980): Wind-stress coefficients over sea surface near neutral conditions—A revisit. *J. Phys. Oceanogr.* **10**, 727–740.

Chapter 20

The Hybrid Parametrical (HYPA) Wave Model

H. Günther and W. Rosenthal

20.1. INTRODUCTION

The Hybrid Parametrical (HYPA) surface wave prediction model for deep water was developed in Sonderforschungsbereich 94 of the University of Hamburg. This model is an extension of the earlier NORSWAM model (Günther *et al.*, 1979a, b; Ewing *et al.*, 1979). It includes, besides parametric prediction of the one-dimensional windsea spectrum, an additional prediction parameter, the mean windsea direction (Günther *et al.*, 1981). The swell propagates on prescribed characteristics for different frequency and direction bins in a way similar to that in the NORSWAM model. In the following sections the three parts of the model (windsea, swell, and windsea–swell exchange) are described.

20.2. WINDSEA

The parametric prognostic equations are deduced from the momentum equation

$$\frac{\partial \mathbf{m}}{\partial t} + (\mathbf{v} \cdot \nabla)\mathbf{m} = \boldsymbol{\tau} \tag{20.1}$$

H. Günther and W. Rosenthal • Institut für Meereskunde der Universität and Max-Planck-Institut für Meteorologie, 2000 Hamburg 13, Federal Republic of Germany.

where $\mathbf{m}$ is the momentum density spectrum, $\mathbf{v}$ the group velocity, and $\boldsymbol{\tau}$ the source function for the spectral momentum density. This is, as usual, divided into the atmospheric input $\boldsymbol{\tau}_{\text{in}}$, the dissipation $\boldsymbol{\tau}_{\text{dis}}$, and the nonlinear interaction $\boldsymbol{\tau}_{\text{nl}}$.

Because it has been shown that the nonlinear interaction in wind-driven surface waves is the main shape-stabilizing mechanism for the energy spectrum (Hasselmann *et al.*, 1973, 1976), we assume a JONSWAP self-similar spectral form, and construct the two-dimensional momentum spectrum as

$$\mathbf{m}(f,\delta) = E_J(f)\cdot\frac{2}{\pi}\cos^2(\delta-\theta_0)\frac{\mathbf{k}}{\omega} \tag{20.2}$$

where

$$E_J(f) = \alpha g^2(2\pi)^{-4}f^{-5}\exp\left\{-\frac{5}{4}\left(\frac{f}{f_{\text{m}}}\right)^{-4} + \ln\gamma\exp\left[-\frac{(f-f_{\text{m}})^2}{2\sigma^2 f_{\text{m}}^2}\right]\right\}$$

is the JONSWAP spectrum containing the free parameters $a_1 = f_{\text{m}}$, $a_2 = \alpha$, $a_3 = \gamma$, $a_4 = \sigma_a$, and $a_5 = \sigma_b$. Furthermore, (20.2) includes the parameter $a_6 = \theta_0$, which is defined as the mean direction of the spectrum averaged over all frequencies.

Mapping (20.1) onto the parameters a_i $(i = 1,\ldots,6)$ of (20.2) leads to the parametric transport equations

$$\frac{\partial a_i}{\partial t} + D_{ijk}\frac{\partial a_j}{\partial x_k} = S_i \qquad \begin{cases} i = 1,\ldots,5 \\ j = 1,\ldots,5 \\ k = 1,2 \end{cases} \tag{20.3}$$

Within this mapping the propagation matrix D_{ijk} is calculated exactly. Due to the complicated structure of the nonlinear interaction source function, the parametric source functions are defined by dimensional analysis for the scale parameters f_{m} and α, and the dependence of the form parameters γ, σ_a, and σ_b is taken from calculations of nonlinear interaction for mean JONSWAP spectra. For the remainder of the source functions, a minimal dissipation (Hasselmann *et al.*, 1976) is assumed. Projecting a Snyder–Cox input term to the parameter space leads to nonzero source functions for the α and θ_0 equations only. Whereas the free coefficient in the α equation is determined from the fetch laws

$$f_{\text{m}}u/g = 2.84\left(x\cdot g/u^2\right)^{-0.3}$$

$$\alpha = 0.0662\left(x\cdot g/u^2\right)^{-0.2} \tag{20.4}$$

which have to be solutions in cases of ideal generation, the free coefficient in the θ_0 equation is fixed from JONSWAP 73 data. Finally, the f_m, γ, and σ_b source functions are slightly modified to achieve a smooth transition to the Pierson–Moskowitz spectrum of a fully developed sea. The α source function constructed in this way contains the component of the wind vector parallel to the mean sea direction, which is used for growing, whereas the θ_0 source function includes the orthogonal component. Therefore, the momentum flux from the atmosphere normal to the sea direction turns the windsea into the new wind direction. Rewriting the θ_0 equation as a relaxation-type equation results in a relaxation time of about 6 h for a wind speed of 20 m/s and a peak frequency $f_m = 0.1$ Hz (Günther *et al.*, 1981).

20.3. SWELL AND WINDSEA–SWELL EXCHANGE

The source function for the α equation becomes zero if $f_m u_{*}/g \leq 0.13$. Therefore, in the model, all momentum with frequencies less than $0.13 \cdot g/u_{*}$ is treated as swell. Momentum in this frequency range is produced when the wind speed decreases or when there are rapid changes in the wind direction. The old windsea is distributed over prescribed characteristics [solutions of (20.1) when $\tau = 0$] , and the energy is propagated freely until it arrives in regions where a windsea with a peak frequency less than the frequency of the swell exists. In this case, the swell energy is removed into the windsea. If swell energy arrives in regions with high wind but only a small developed windsea, a Snyder–Cox-type amplification is applied.

20.4. APPLICATION TO WAVE MODEL INTERCOMPARISON CASES

The above-described model was applied to the wave model intercomparison cases. The runs were carried out with a four-parameter version of the model; that means the peak width parameters σ_a and σ_b of the windsea are not predicted by the corresponding equation but defined as the solutions of the source functions $S_5 = S_6 = 0$. This seems to be possible because the relaxation times for these parameters to their equilibrium are much smaller than the scales corresponding to the other parameters and to the resolution in space and time.

In most cases of the intercomparison, the models should be initialized with a calm sea, and there should be no energy flux into the grid across the boundaries. Because the present model uses the peak frequency as a prediction parameter and zero energy corresponds to $f_m = \infty$, we initialized the runs with a Pierson–Moskowitz spectrum of $f_m = 0.4$ Hz. If boundary values are required, we use fetch laws with a fetch equal to the grid size, which is 25 km in most of the cases. These modifications will not influence the stationary solutions, but if we are looking for time-dependent development the model output will be slightly biased to earlier times.

REFERENCES

Ewing, J. A., T. J. Weare, and B. A. Worthington (1979): A hindcast study of extreme wave conditions in the North Sea. *J. Geophys. Res.* **84**, 5739–5747.

Günther, H., W. Rosenthal, T. J. Weare, B. A. Worthington, K. Hasselmann, and J. A. Ewing (1979a): A hybrid parametrical wave prediction model. *J. Geophys. Res.* **84**, 5727–5738.

Günther, H., W. Rosenthal, and K. Richter (1979b): Application of the parametrical surface wave prediction model to rapidly varying wind-fields during JONSWAP 1973. *J. Geophys. Res.* **84**, 4855–4864.

Günther, H., W. Rosenthal, and M. Dunckel (1981): The response of surface gravity waves to changing wind directions. *J. Phys. Oceanogr.* **11**, 718–728.

Hasselmann, K., T. P. Barnett, E. Bouws, H. Carlson, D. E. Cartwright, K. Enke, J. A. Ewing, H. Gienapp, D. E. Hasselmann, P. Kruseman, A. Meerburg, P. Müller, D. J. Olbers, K. Richter, W. Sell, and H. Walden (1973): Measurements of wind-wave growth and swell decay during the Joint North Sea Wave Project (JONSWAP). *Dtsch. Hydrogr. Z., Suppl. A* **8** (12).

Hasselmann, K., D. B. Ross, P. Müller, and W. Sell (1976): A parametrical wave prediction model. *J. Phys. Oceanogr.* **6**, 201–228.

Chapter 21

The UK Meteorological Office Operational Wave Model (BMO)

B. Golding

The operational wave prediction model used by the UK Meteorological Office has a grid point representation of the wave field. This is usually defined on a polar stereographic projection of the earth's surface, and it can be initialized with any grid length and any shape or size of ocean basin, subject to core storage constraints. It has a general nesting capability, permitting any model to obtain inflow boundary conditions from any other. Operationally, two models are run. One covers much of the Northern Hemisphere, with a grid length at 60°N of 300 km. This provides boundary conditions for a 50-km-grid-length model of the Northwest European Continental Shelf. An extended description of the model is given in Golding (1983).

The wave field at each point is described by a discrete spectrum. The spectral resolution is variable but at present comprises 12 direction components (30° resolution) and 11 frequencies (0.05, 0.06, 0.072, 0.086, 0.104, 0.124, 0.148, 0.178, 0.214, 0.256, 0.308 Hz). Shallow-water effects are included as an option. They are calculated with a 2-m depth resolution and model the influence of variable group velocity, refraction, and bottom friction. They have a significant effect on the results in the North Sea and English Channel.

The model solves the energy balance equation

$$\frac{dE}{dt} = f(E)$$

B. Golding • British Meteorological Office, Bracknell, Berkshire RG12 2SZ, United Kingdom.

starting from initial conditions, which may be zero or prescribed by the results of the last integration, and subject to radiative boundary conditions with zero inflow at coasts and prescribed inflow at open sea boundaries. The balance equation is actually solved in the following form:

$$\left[\left[\left[\left[\frac{\partial}{\partial t}(E_{ij}) = -\nabla\cdot(\mathbf{c}_g E_{ij})\right]_{\text{(i)}} - \frac{\partial}{\partial\theta}\left\{(\mathbf{c}_g\cdot\nabla\theta)E_{ij}\right\}_{\text{(ii)}}\right] + \alpha + \beta \mathrm{E_{ij}} - \Delta(E)\right]_{\text{(iii)}} + \Gamma(E)\right]_{\text{(iv)}}$$

Multiplicative splitting is used so that integration of each numbered section uses the results of the previous section. An attempt has been made above to indicate those parts which operate on the uncoupled spectral components (E_{ij}) and those which operate on the whole spectrum (E). The detailed form of each term is given below. A variable time step is used in the model. The basic time step is equal to the time interval between successive updates of the wind field. Divisors and multiples of this time step are used in sections (i), (ii) such that the stability criterion for the propagation scheme is always satisfied. In practice this means that the ratio of gridlength to timestep varies between about 0.4 and 0.8 of the deep water group velocity.

i. Propagation

$$\frac{\partial}{\partial t}(E_{ij}) = -\nabla\cdot(\mathbf{c}_g E_{ij})$$

This term is integrated using a modified Lax–Wendroff integration scheme. For one-dimensional propagation at constant speed c it takes the form

$$E_{j+\frac{1}{2}}^{n+\frac{1}{2}} = \tfrac{1}{2}\left(E_j^n + E_{j+1}^n\right) - \frac{\mu}{2}\left(E_{j+1}^n - E_j^n\right)$$

$$E_j^{n+1} = E_j^n - \mu\left[(1+a)\left(E_{j+\frac{1}{2}}^{n+\frac{1}{2}} - E_{j-\frac{1}{2}}^{n+\frac{1}{2}}\right) - \frac{a}{3}\left(E_{j+\frac{3}{2}}^{n+\frac{1}{2}} - E_{j-\frac{3}{2}}^{n+\frac{1}{2}}\right)\right]$$

where $\mu = c(\Delta t/\Delta x)$, $a = \frac{3}{4}(1-\mu^2)$, and j signifies spatial position and n the time level. For details of its derivation see Gadd (1978) and for its application to the wave model see Golding (1977).

ii. Refraction

$$\frac{\partial}{\partial t}(E_{ij}) = -\frac{\partial}{\partial\theta}\left[(\mathbf{c}_g\cdot\nabla\theta)E_{ij}\right]$$

The refraction scheme is based on Snell's law for optical refraction and details of the derivation are given in Golding (1977). The resultant form for $\mathbf{c}_g \cdot \nabla\theta$ may be written

$$\mathbf{c}_g \cdot \nabla\theta = -\frac{|\mathbf{c}_g|}{k}\left(\frac{\partial H}{\partial x}\sin\theta - \frac{\partial H}{\partial y}\cos\theta\right)\left(\frac{-k^2\operatorname{sech}^2 kH}{\tanh kH + kH\operatorname{sech}^2 kH}\right)$$

where H is water depth and k is wavenumber. The numerical scheme uses centered differences in θ.

iii. Growth and Decay

$$\frac{\partial}{\partial t}(E_{ij}) = \alpha + \beta E_{ij} - \Delta(E) - \Phi(E)$$

The α term initiates wave growth but is unimportant thereafter. A simple form is used which removes the effects of variable spectral resolution:

$$\alpha = \begin{cases} \alpha_1 U^2\cos^2(\theta - \psi) & \text{for highest frequency if } |\theta - \psi| \le 90^\circ \\ 0 & \text{otherwise} \end{cases}$$

where U and ψ are the wind speed and direction.

The β term gives exponential growth. It has a cutoff for waves traveling faster than the wind.

$$\beta = \begin{cases} \beta_1 f\left[\dfrac{U\cos(\theta - \psi)}{v_f} - 1\right] & \text{for } \dfrac{U\cos(\theta - \psi)}{v_f} > 1 \\ 0 & \text{otherwise} \end{cases}$$

where $v_f = g/2\pi f$ is the phase speed of the waves.

Dissipation is based on Hasselmann's whitecapping model (1974):

$$\Delta(E) = \Delta_1 f^2 \bar{E}^{0.25} E_{ij}$$

where $\bar{E}$ is the total spectral energy. Values of $\alpha_1(6\times10^{-8}/2\pi f_{\max})$, $\beta_1(6\times 10^{-2}\times 2\pi\rho)$, and $\Delta_1(4\times10^{-4})$ were chosen to reproduce a duration-limited growth curve based on the JONSWAP results.

Bottom friction is based on Collins' formulation for a quadratic friction term (Collins, 1972):

$$\Phi(E) = \frac{\Phi_1 g k^2 |\mathbf{c}_g|}{(2\pi f\cosh kH)^2}\left[\int_f\int_\theta\left(\frac{gk}{2\pi f\cosh kH}\right)^2 E(f,\theta)\,d\theta\,df\right]^{0.5} E(f,\theta)$$

where $\Phi_1(=0.05)$ was chosen to give results consistent with those of Bretschneider (1954) for wave generation in shallow water.

iv. Nonlinear Interactions. The windsea part of the spectrum is first separated from the rest. This is done by assuming that windsea energy will lie within 90° of the wind direction and at frequencies greater than 0.8 of the peak frequency of the local Pierson–Moskowitz spectrum. A peak frequency is calculated appropriate to the energy in this region of the spectrum, and the windsea is then defined as energy within 90° of the wind direction and greater than 0.8 of this peak frequency.

It is then assumed that the nonlinear interactions will act to transform the windsea into a JONSWAP spectrum directionally centered on the wind direction with peak frequency and peak enhancement uniquely related to its energy. The spectrum is transformed to this shape with its total energy conserved. The transformation is immediate and couples all components with frequency greater than 0.8 of the peak and direction within 90° of the wind.

Wind input is obtained from the operational numerical atmospheric prediction system of the UK Meteorological Office. For the diagnosis of the present sea state a number of sources of information are used. The most recent forecast for the past 12 h forms the basis. This is corrected at the end by winds derived from the operational pressure analysis. Then, it is corrected every 3 h by insertion of wind observations using a two-parameter orthogonal polynomial analysis. Finally, the corrections are applied in linearly decreasing amounts away from the observation times. The forecast and analysis values are obtained from the 900-mb fields by applying formulas derived from Findlater *et al.* (1966):

$$U = aU_{900}^2 + bU_{900} + c$$

$$\psi = \psi_{900} + d$$

where a, b, c, and d are stability-dependent constants. These equations provide 19.5-m winds. No corrections for height are made to the observations.

A substantial program of routine verification is maintained. Typical results from the comparison with buoy reports are shown in the table below. It shows RMS wave height errors in meters at two locations for the initial and 24-h forecast fields during the period June 1980–May 1981.

	$T+0$	$T+24$
Southern North Sea	0.5	0.6
Southwest approaches	0.9	1.0

Since it is an operational model, the formulation is constantly under review. The description given above applies to the version in use from spring 1980 to Autumn 1982.

REFERENCES

Bretschneider, C. L. (1954): Generation of wind waves over a shallow bottom. Technical Memorandum 51, Corps of Engineers Beach Erosion Board, Department of the Army, Washington, D.C.

Collins, J. (1972): Prediction of shallow water spectra. *J. Geophys. Res.* **77,** 2693–2707.

Findlater, J., T. N. S. Harrower, G. A. Howkins, and H. L. Wright (1966): Surface and 900 mb wind relationships. Meteorological Office Science Paper 23, HMSO, London.

Gadd, A. J. (1978): A numerical advection scheme with small phase speed errors. *Q. J. R. Meteorol. Soc.* **104,** 583–594.

Golding, B. W. (1977): A depth dependent wave model for operational forecasting. *Turbulent Fluxes through the Sea Surface, Wave Dynamics, and Prediction* (A. Favre and K. Hasselmann, eds.), Plenum Press, New York. 593–604.

Golding, B. W. (1983): A wave prediction system for real-time sea state forecasting. *Q. J. R. Meteorol. Soc.* **109**, 393–416.

Hasselmann, K. (1974): On the spectral dissipation of ocean waves due to whitecapping. *Boundary-Layer Meteorol.* **6,** 107–127.

Chapter 22

Intercomparison Test Version of the SAIL Wave Model

J. A. Greenwood, V. J. Cardone, and L. M. Lawson

22.1. INTRODUCTION

In this chapter, a new wave prediction model termed discrete/hybrid, is proposed and tested, which incorporates parts of both Pierson–Tick–Baer (PTB)-type models and parametric models. It differs from the hybrid/parametric models in that the entire directional spectrum is banded and all bands are propagated. It differs from the PTB treatment of the source term in applying in banded space a parametric growth to model the effects of nonlinear wave–wave interactions. In the new model, each band is subjected to the greater of growth by a wind-wave instability mechanism consistent with the Abaco measurements, and parametric growth derived from an $\varepsilon-\xi$ law (ε is nondimensional total energy; ξ is nondimensional fetch).

Unlike the PTB growth model, frequency bands are coupled by basing parametric growth on ε_1 (nondimensional energy computed over downwind bands) and basing equilibrium range relaxation, which also redistributes energy angularly, on an $\alpha-\varepsilon_2$ law (ε_2 is nondimensional total energy). An explicit directional relaxation of the windsea is also included.

The growth of wave energy with fetch and duration of steady winds is tuned through the $\varepsilon_1-\xi$ and $\alpha-\varepsilon_2$ laws. In a turning wind, the time scale of the directional relaxation also influences growth. This chapter describes the particular tuning adopted for the intercomparison tests.

J. A. Greenwood and V. J. Cardone • Oceanweather Inc., Cos Cob, Connecticut 06807, USA.
L. M. Lawson • Atlantic Oceanographic and Meteorological Laboratories, National Oceanic and Atmospheric Administration, Miami, Florida 33149, USA.

PTB models are represented in the intercomparison by the MRI model; see Chapter 15.

22.2. MODEL STRUCTURE

22.2.1. Spectral Bands

The operational version and most tested versions of the model have 24 directional bands spaced 15° apart. The first directional band is 7.5° from grid north since it was found that for our propagation system bands propagating along grid lines tended to produce dead areas that swells could not reach.

The intercomparison test model had 15 frequency bands, from 0.04 Hz to 0.22093 Hz. The ratio between adjacent bands is $3^{1/9}$. The highest band is considered to contain the entire tail of the spectrum from 0.20785 Hz to infinity. To the extent that spectra are similar, this spacing gives a resolution independent of wind speed. That is, by spacing bands in geometric progression to the extent that spectra are similar, the number of bands between half-power points of the spectral density is independent of peak frequency.

22.2.2. Grid Systems

The grid used in production runs of the model is a square grid laid out on either a direct mercator projection or a transverse mercator projection; ellipticity of the earth is ignored, and so the two cases yield the same propagation formulas. The propagation scheme discussed below has also been applied on a rectangular grid defined by meridians at one spacing and parallels at the same or different spacing, on an isosceles triangular grid formed by staggering points in alternate parallels, and on an isosceles triangular grid formed from a square mercator grid by retaining only points where $I + J$ is an even integer. In preparing to run hindcasts for any basin, the first step is to compute the true latitude and longitude of grid points and to plot them on the scale of an available marine chart so that land points can be identified and excluded from the points from which energy is propagated.

For Cases III–VI of the intercomparison study a square grid was laid out on a flat earth.

For Case II, and for fetch-limited tests carried out for our internal use, a one-dimensional grid was used. On a flat earth, one line is erected perpendicular to a straight coast, and grid points are arrayed on that line at spacing Δx. Spectra at all points along a line parallel to the coast are treated as identical to the spectrum at the one representative point that is carried in the computer; when establishing the computational form of the propagation formula, points with the same fetch and different along-shore coordinates are combined. It is thus possible on a moderate-sized computer to run a fetch-limited test in core for 180 or more fetches and either orthogonal or oblique winds.

22.2.3. Wind Scaling

There remains considerable uncertainty over whether friction velocity u_*, or the wind at anemometer height, U_z, is the more appropriate velocity scale for wave prediction models. Most modelers use U_{10} or $U_{19.5}$, but recent field evidence offered by Liu and Ross (1980) and Forristall (1981) supports the use of u_* to scale wind–wave relationships. The present model obeys similarity of wave growth with fetch and duration when U_{10} is used consistently. However, we have adopted u_* scaling for parametric growth and direct (B-term) growth and U_{10} scaling for the Pierson–Moskowitz (PM) fully developed spectrum and saturation value of ε. Since the ratio of u_*/U_{10} is not constant, on the evidence of recent field data (Large and Pond, 1981) our model is not similar to either u_* or U_{10}.

The drag law connecting u_* and U_{10} is

$$U_z = \frac{u_*}{k} \log_e \frac{z}{z_0}$$

where $k = 0.35$ and $z_0 = \max(3.15 \times 10^{-4}, 0.035 u_*^2/g)$, which is a modification of Garratt (1977) in accordance with the measurements of Large and Pond. This scaling law yields $u_* = 0.846$ m/s when the wind at 10 m is 20 m/s. However, in Cases II–V, a u_* of 0.855 m/s was used as agreed.

22.2.4. Numerical Solution

The model works by alternate growth and propagation steps. In a growth step, the several grid points are totally uncoupled; the frequency and direction bins at one grid point are weakly coupled because of the equilibrium-range relaxation and directional relaxation discussed below. In a propagation step, the frequency bands are totally uncoupled; the directional bands are weakly coupled by convergence of meridians on a spherical earth. In flat-earth tests, such as Cases III–VI, the directional bands are uncoupled in the propagation step.

In order to approximate the effects of simultaneous growth and advection each time step of length Δt is divided into three parts: (1) grow for

$\frac{1}{2}\Delta t$; (2) propagate for Δt; and (3) grow for $\frac{1}{2}\Delta t$ again. This cycle was devised for medium-sized computers (where spectra on five parallels can be held in core at propagate time) and adapted for small computers (where spectra at the points adjoining the current point must be fetched from disk before propagating). On computers where the spectra at all points in a basin will fit in core, it is advantageous to invert the cycle, viz., propagate for $\frac{1}{2}\Delta t$; grow for Δt; propagate for $\frac{1}{2}\Delta t$ again.

22.3. PROPAGATION SCHEME

The propagation scheme used was constructed for use with a spherical earth and combines elements of jump and interpolatory propagation. When an ocean basin is mapped on a plane by an arbitrary projection and a rectangular or triangular grid overlaid, the distance from each point to its neighbors, and thus the coefficients in the propagation formula, are functions of both latitude and longitude. Coefficients dependent on latitude alone arise when one set of grid lines is meridians equally spaced and the other set parallels at any convenient spacing; if the earth is treated as flat and the parallels are equally spaced, then the coefficients are independent of longitude as well.

When using a square grid, we adopt a time step such that $c_g \Delta t < 2.017\Delta x_{\min}$, where c_g is computed for the lowest frequency being modeled (usually 0.04 Hz). The propagation pattern is then the interior of an octagon bounded by the lines $|x_0 - x_1| = 2\Delta x$, $|y_0 - y_1| = 2\Delta x$, and $|x_0 - x_1| + |y_0 - y_1| = 3\Delta x$; the octagon is divided into 28 isosceles triangles by the lines

$$|x_0 - x_1| = 0, \Delta x$$

$$|y_0 - y_1| = 0, \Delta x$$

$$|x_0 - x_1| + |y_0 - y_1| = \Delta x, 2\Delta x$$

Interpolation is performed between the three vertices of whatever triangle the point $(c_g \Delta t \sin\theta, c_g \Delta t \cos\theta)$ falls in; thus propagation is strictly interpolatory for $C_g \Delta t(|\sin\theta| + |\cos\theta|) < \Delta x$, and incorporates jumps for larger c_g.

The mean square dispersion after n time steps is

$$S^2 = An\Delta x^2 + B(nc_g \Delta t)^2$$

where A and B are irregularly varying functions of θ and c_g; with bands spaced as stated above, A has a maximum of $\frac{1}{2}$ and an average value of $\frac{1}{3}$, and B has a maximum of 0.0208 and an average value of 0.0138.

22.4. SPECTRAL GROWTH

22.4.1. Notation

E Mean square instantaneous height. $E = S(\infty)$.
f Frequency, in hertz.
g Acceleration of gravity.
s Exponent in Mitsuyasu's spreading function.
S Integrated spectrum (*not* spectral density).
u_* Friction velocity.
U_{10} Effective wind speed (reduced to neutral temperature profile) at 10-m height.
x Geometrical fetch (idealized, perpendicular to straight coast).
z_0 Roughness length.
α Phillips number $= \lim_{\omega \to \infty} 5g^{-2}\omega^4[E - S(\omega)]$.
γ Peak enhancement $= \alpha^{-1}\hat{\omega}^5 g^{-2} e^{1.25}\left[\dfrac{ds}{d\omega}\right]_{\omega=\hat{\omega}}$.
Γ Integral enhancement: $1 + \Gamma = 5\alpha^{-1}\varepsilon\nu^4 = 5(2\pi)^4\lambda$.
Δt Time step.
$\Delta\tau$ Nondimensional time step $= gU_{10}^{-1}\,\Delta t$.
ε Nondimensional mean square height $= Eg^2U_{10}^{-4}$.
λ Hasselmann's shape parameter $= \alpha^{-1}(\nu/2\pi)^4\varepsilon$.
ν Nondimensional peak frequency $= \hat{\omega}U_{10}/g$. (Note that Hasselmann's ν is our $\nu/2\pi$.)
ξ Nondimensional fetch $= xg/U_{10}^2$.
Φ Coefficient in Phillips' form of ε–ξ law, i.e., $\varepsilon = \Phi\psi\xi$. According to Phillips, $\Phi = 1.6 \times 10^{-4}$; in the 1980 model, $\Phi = 1.3 \times 10^{-4}$ is used.
ψ Drag coefficient $= (u_*/U_{10})^2$.
ω Frequency, radians/s, i.e., $\omega = 2\pi f$.
$\hat{\omega}$ Peak frequency.
Ω Nondimensional frequency $= \omega U_{10}/g$.

22.4.2. Reference Spectrum

We take the PM spectrum in the form

$$S = 0.2\alpha\hat{\omega}^{-4}g^2\exp(-1.25\hat{\omega}^4\omega^{-4})$$

where $\varepsilon = 0.2\alpha\hat{\omega}^{-4}(g/U_{10})^4 = 0.00364$ exactly, and $\alpha = \varepsilon^{-0.23}/449$ exactly; whence $\alpha = 8.104017 \times 10^{-3}$, and $\nu/2\pi = 0.1300102$. For the reference spectrum of a growing sea we adopt the form

$$S = \frac{\varepsilon g^{-2} U_{10}^4}{1+\Gamma}\left[\exp(-1.25\hat{\omega}^4\omega^{-4}) + \Gamma\exp(-1.0625\hat{\omega}^{16}\omega^{-16})\right] \quad (22.1)$$

where $\Gamma = \max\{0, 0.54[1-(\varepsilon/0.00364)^2]\}$, $\alpha = \varepsilon^{-0.23}/449$, and $\hat{\omega}^4 = (g/U_{10})^4\varepsilon^{-1.23}(1+\Gamma)/2245$. The peak enhancement γ varies from 3.214 at $\varepsilon = 0$ to 1 at $\varepsilon \geq 0.00364$. In practice, because the frequency bands used cannot accurately resolve a narrow peak, values of γ greater than 2.9 are not observed. Our angular spreading function is essentially Mitsuyasu's, modified to be rigorously zero in the upwind quadrants:

$$M(\theta) = \begin{cases} 0 & \cos(\theta - \theta_\omega) \leq 0 \\ I^{-1}\{[1+\cos(\theta-\theta_\omega)]^s - 1\} & \cos(\theta - \theta_\omega) > 0 \end{cases} \quad (22.2)$$

where

$$s = \min\left[11.5(\omega U_{10}/g)^{-2.5}, 52.4226(\omega U_{10}/g)^5\right]$$

and

$$I = -\pi + \int_{-1/2\pi}^{+1/2\pi}\left[1+\cos(\theta-\theta_\omega)\right]^s d\theta$$

22.4.3. Growth on Forward Face

In each frequency–direction bin, B-term growth is computed as $\Delta F = \beta\omega F\Delta t$, where F is the component of mean square elevation in a bin, and

$$\beta = \max\left[0, 3\times 10^{-4}(30u_*\omega g^{-1}\cos\theta - 1)\right]$$

where θ is the angle between the wind and the component of waves.

Parametric growth is referred to the single parameter ε_1 (a weighted variant of ε, computed over downwind bands). The spectrum (22.1) is transformed into the computational form

$$S = g^{-2}U_{10}^4\varepsilon\left\{(1+\Gamma)^{-1}\exp(-1.25\nu^4\Omega^{-4}) + \Gamma(1+\Gamma)^{-1}\exp\left[-1.0625(\nu^4\Omega^{-4})^4\right]\right\} \quad (22.3)$$

where $\Omega = \omega U_{10}/g$, $\bar{\Gamma}$ is computed by (22.2), and

$$\nu^4 = \Gamma/(2245\varepsilon^{1.23})$$

The parametric growth in the frequency band (ω_1, ω_2) is given by $\Delta E = S(\omega_2, \varepsilon_2) - S(\omega_1, \varepsilon_2) - S(\omega_2, \varepsilon_1) + S(\omega_1, \varepsilon_1)$, where

$$\varepsilon_2 = \varepsilon_1 + \Phi\psi\Delta\xi \qquad \text{and} \qquad \Delta\xi = \tfrac{1}{2}g\bar{\omega}^{-1}\Delta t \times gU_{10}^{-2} \qquad (22.4)$$

For each frequency–direction bin the adopted growth is the greatest of zero, B-term growth, and ΔE, spread over directional bands by a $\cos^3$ law. (This is the equivalent, on the time side, of $\cos^2$ law used by Hasselmann on the fetch side.) In (22.4), $\omega = (\omega_1\omega_2)^{1/2}$, with appropriate modifications in the lowest and highest frequency bands. For frequencies less than the PM peak, if the energy in a band exceeds PM it is not grown; and if "adopted growth" would take a band above PM, it is grown to PM at which point growth abruptly stops.

22.4.4. Equilibrium Range Relaxation

The previous step has applied B-term growth to the right of the local peak frequency. (In this range parametric growth is typically negative and is

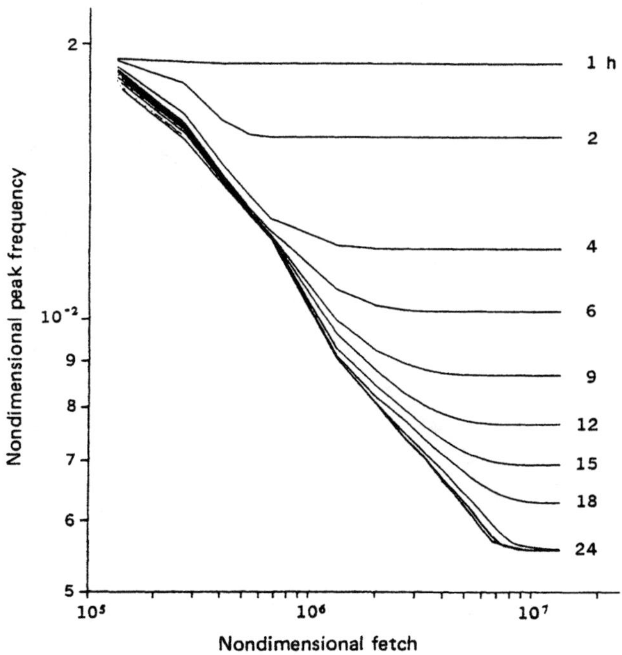

Fig. 22.1. Nondimensional peak frequency $(u_* f_p g^{-1})$ versus nondimensional fetch $(\times gu_*^{-2})$ and duration from Case II.

not applied.) To restore the tail of the spectrum to an $\alpha g^2\omega^{-5}$ form, ε is next computed from the variance E summed over all frequencies and all directions; the presence of a heavy swell can inflate ε above 0.00364 and depress α below 0.0081. From this ε is computed an imputed peak frequency

$$\hat{\omega} = gU_{10}^{-1}\left[\varepsilon^{-1.23}(1+\Gamma)/2245\right]^{0.25}$$

For frequency bands to the right of $\hat{\omega}$ *and* to the right of the PM peak, the excess of the variance in a frequency band over that given by (22.1) is removed, using a weighting function that is 1 downwind and $1-\cos(\theta-\theta_\omega)$ upwind. The effect of this step is to enforce an $\alpha-\varepsilon$ relation in the equilibrium range. For $\omega > \hat{\omega}$, this step implicitly decays wave components moving upwind; for $\omega < \hat{\omega}$, upwind components are not decayed, and are removed only by propagating elsewhere. (Thus in Case VII, low-frequency swells persist undiminished to the end of the test.)

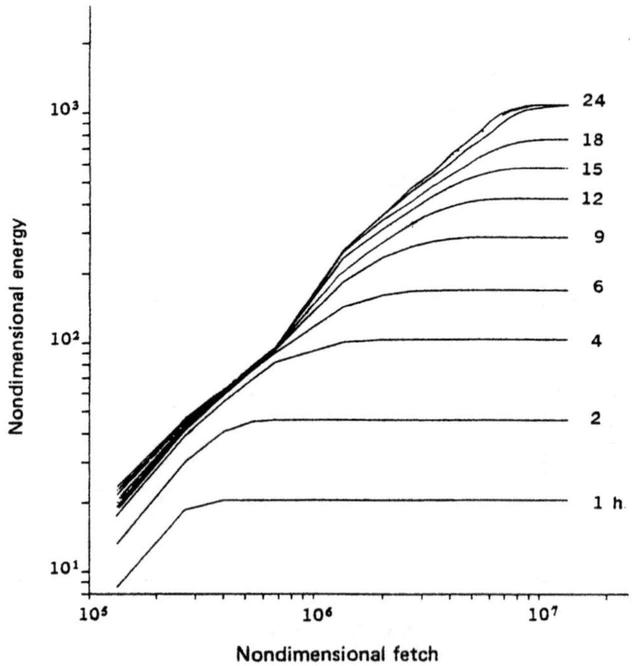

Fig. 22.2. Nondimensional energy ($Eg^2 u_*^{-4}$) versus nondimensional fetch and duration from Case II.

22.4.5. Directional Relaxation

The ideas in the computation below are that bands in the equilibrium range are to be turned into equilibrium with the wind direction; bands on the forward face are to undergo a partial relaxation toward the wind direction; and bands to the left of the PM peak are considered swell and left unaltered. For each band to the right of the PM peak, find H, the largest energy in an upwind directional bin. Frequency–direction bins with energy less than the reference spectrum (PM with Mitsuyasu's spreading) are eligible to be pulled up to the reference spectrum; bins greater than the reference spectrum are eligible to be pulled down to the reference spectrum, but not below H. (The effect of the last clause is that, in a turning wind, the band 82.5° off the wind decays not faster than the band 97.5° off the wind.) In frequency bands where the energy exceeds the PM spectrum (i.e., the saturated tail), if the eligible increases and decreases balance, they are all made instantly; if the eligible increases exceed the eligible decreases, all of the decreases are made, as well as the fraction of the increases needed to

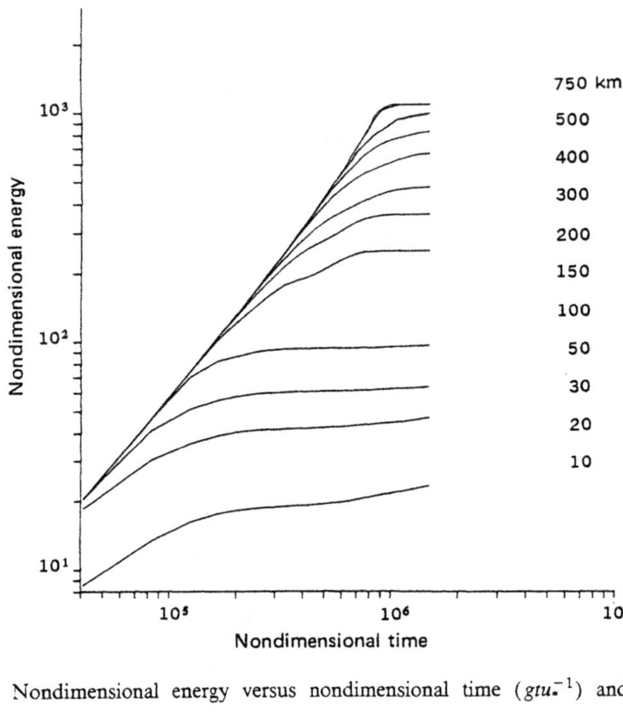

Fig. 22.3. Nondimensional energy versus nondimensional time (gtu_*^{-1}) and fetch from Case II.

maintain the energy in the frequency band unchanged; if the eligible increases fall short of the eligible decreases, all the increases and the appropriate decreases are made. In frequency bands where the energy does not exceed the PM spectrum (i.e., the forward face) tentative increases and decreases are computed as in the previous sentence, but only a fraction of this redistribution is effected: this fraction equals the energy in the band divided by the PM spectrum for the same band.

22.5. GROWTH CHARACTERISTICS

The principal growth characteristics of the new model are seen in the results for Case II, which were computed for a U_{10} of 20 m/s. Figures 22.1–22.4 show the growth of energy and peak frequency (nondimensionalized) with nondimensional fetch and duration. The development of

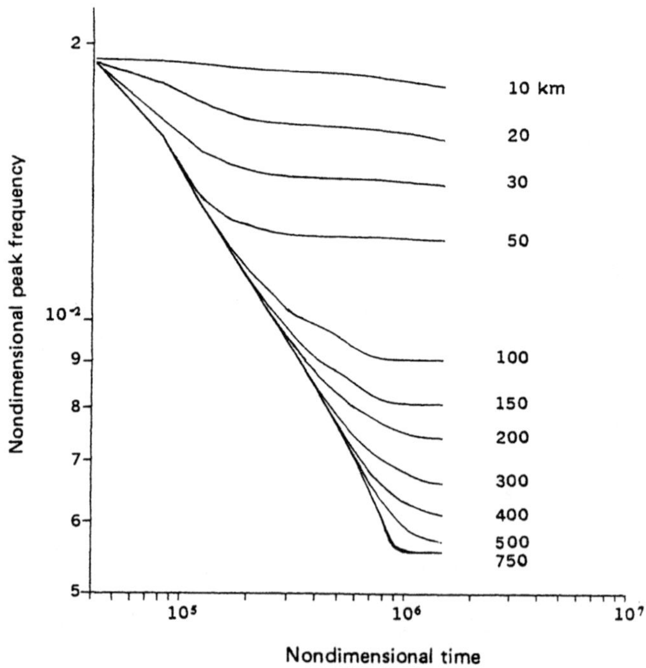

Fig. 22.4. Nondimensional peak frequency versus nondimensional time and fetch from Case II.

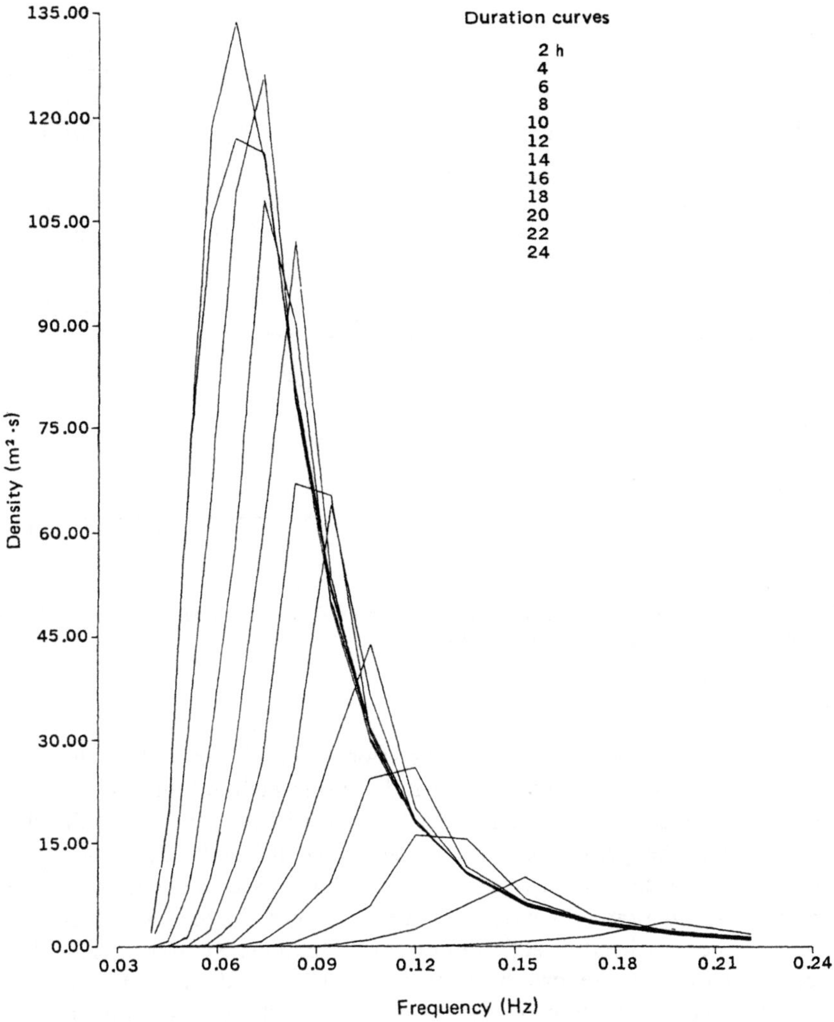

Fig. 22.5. Growth of the spectrum with duration in Case II.

the frequency spectrum with duration for 20-m/s winds is shown in dimensional form in Fig. 22.5. For these runs the coefficient ϕ in the $\varepsilon_1-\xi$ law was fixed at 1.3×10^{-4}, which provides growth rates similar to those of the ODGP model of Cardone *et al.* (1976).

Figures 22.6 and 22.7, from Case VII, show the variation of total energy and the time it takes for the mean direction to turn 45° (T_{45})

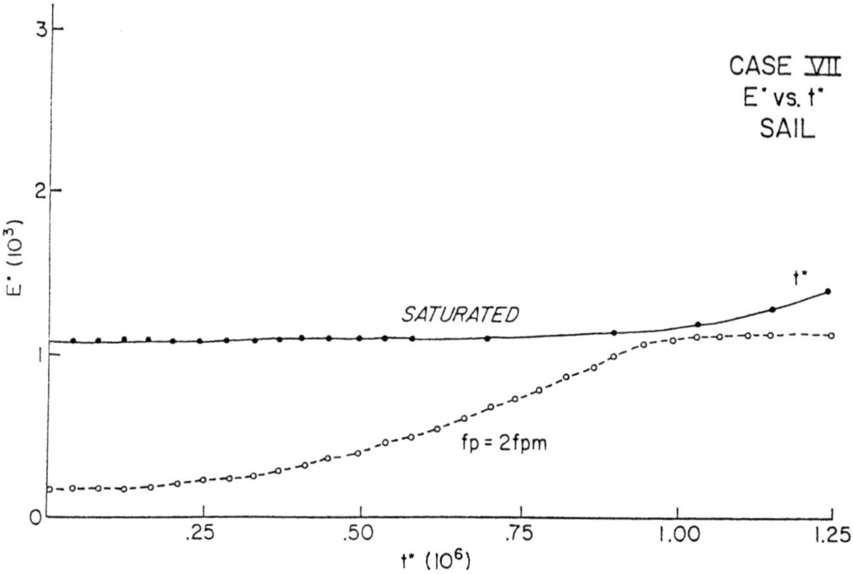

Fig. 22.6. Nondimensional energy versus nondimensional time following an abrupt wind shift of 90° for two initial conditions of Case VII.

following an abrupt shift of wind direction from 90°. In one of the cases shown the shift occurs after a sufficient duration (fetch-unlimited) of 20 m/s winds has elapsed to achieve full development; in the other case, the shift occurs when the peak frequency equals twice the fully developed peak frequency. The directional relaxation algorithm produces a relaxation time which is a function of windsea frequency. Lower-frequency components adjust to the new wind direction more slowly than higher frequencies. More recent versions of the proposed model included a directional-relaxation algorithm with a tunable relaxation time, since the growth properties in turning winds were found to be very sensitive to this mechanism.

Since the intercomparison tests were run, the model has continued to undergo considerable testing against wave measurements made in real tropical and extratropical storms. Those results will be reported in a separate paper. Apart from the modification of the directional relaxation (tuned to provide a longer relaxation time than for the intercomparison version), the principal change required to achieve accurate hindcasts was to slow the growth rate of total energy in the $\varepsilon_1 - \xi$ law as full development was approached.

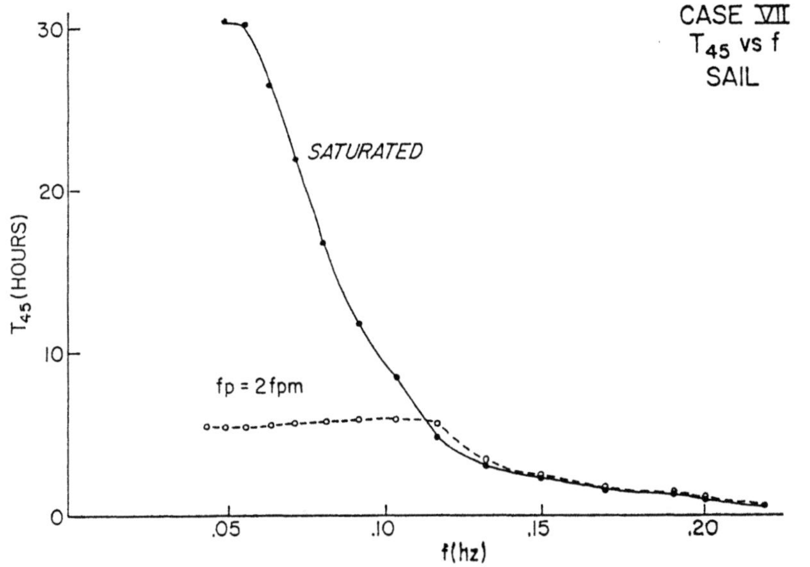

Fig. 22.7. Time for mean direction to turn 45° (T_{45}) versus frequency for two initial conditions of Case VII.

REFERENCES

Cardone, V. J., W. J. Pierson, Jr., and E. G. Ward (1976): Hindcasting the directional spectrum of hurricane-generated waves. *J. Pet. Technol.* **28,** 385–394; *Trans. AIME* **261.**

Forristall, G. Z. (1981): Measurements of a saturated range in ocean wave spectra. *J. Geophys. Res.* **86,** 8075–8084.

Garratt, J. R. (1977): Review of drag coefficients over ocean and continents. *Mon. Weather Rev.* **105,** 915–929.

Large, W. G., and S. Pond (1981): Open ocean momentum flux measurements in moderate to strong winds. *J. Phys. Oceanogr.* **11,** 324–336.

Liu, P. C., and D. B. Ross (1980): Airborne measurements of wave growth for stable and unstable atmospheres in Lake Michigan. *J. Phys. Oceanogr.* **10,** 1842–1853.

Chapter 23

The DNS Model: An Improved Spectral Model for Ocean Wave Prediction

J. H. Allender, T. P. Barnett, and M. Lybanon

23.1. INTRODUCTION

Any attempt to model the frequency- and direction-dependent ocean wave spectrum $F(f, \theta)$ is a formidable task. The only more or less unified theory (Hasselmann, 1967) has not been reduced to a practical form. The approach taken in this study is to combine individual, existing theories to produce results that are compatible with our present understanding of the evolving wind wave spectrum. In this chapter we discuss a discrete spectral model that uses an adaptation of the method of characteristics to propagate energy. The numerical method was discussed originally by Barnett *et al.* (1969).

We begin with a description of the way that physical processes affecting the development of the wave spectrum are represented. In general the present model physics is better than that in previous spectral models. Additional logic governing the spectral balance in particular parts of the (f, θ) plane is then outlined. We define the discrete equations and give an explanation of the numerical method, which is quite general and gives accurate propagation. Key aspects of model behavior are reviewed with regard to the ideal cases of wave generation from the model intercompari-

J. H. Allender • Naval Ocean Research and Development Activity, NSTL Station, Mississippi 39529, USA. *Present address:* Chevron Oil Field Research Company, La Habra, California 90631, USA. **T. P. Barnett** • Scripps Institution of Oceanography, La Jolla, California 92093, USA. **M. Lybanon** • Naval Ocean Research and Development Activity, NSTL Station, Mississippi 39529, USA.

son study. We find that the model imitates the JONSWAP results for fetch- and duration-limited wave growth rather well. Finally, model capabilities and future improvements are discussed.

23.2. REPRESENTATION OF PHYSICAL PROCESSES

Numerous physical processes affect the evolving wave spectrum. For the present model we assume that input from the atmosphere (S_{in}) and weak, nonlinear, wave–wave interactions (S_{nl}) govern the growth of the wave spectrum. Dissipation, i.e., any processes that remove energy from the surface wave field, enters only through a prescribed, maximum spectral density that depends on the local wind speed. The resulting combination of source terms agrees with that proposed by Hasselmann *et al.* (1976), and, just as importantly, imitates crudely the best measurements of the net source term for growing seas (Hasselmann *et al.*, 1973).

23.2.1. Atmospheric Input

The form of S_{in} yields an exponential growth mechanism, consistent with linear feedback theories for atmospheric input. The two-dimensional (2d) source function takes the form:

$$S_{in}(U,f,\theta) = \beta F(f,\theta)$$

$$\beta = \begin{cases} 1.51(10^{-3})(2\pi fU/g\cos\theta_I - 0.9)f \geq 0 & \text{for } \theta_I = |\theta - \theta_U| \leq \pi/2 \\ 0 & \text{for } \theta_I > \pi/2 \end{cases}$$

(see Appendix for definition of symbols). The growth coefficient β has been adjusted to give the minimum momentum transfer found by Snyder *et al.* (1981). This coefficient is one-fourth the value used in Barnett (1968); the nonlinear source term (described below) makes up the difference in the present model. The sharp cutoff in wave generation when the wind speed is 0.9 times the phase speed of a given wave component is unrealistic. In the model we assume the wind speed is normally distributed about a mean speed with a standard deviation of 10%. The resulting β value is an average based on this distribution and wind speeds spanning two standard deviations about the mean. This procedure makes for a smoother transition out of the range of frequencies where active generation takes place. Note that the growth coefficient is nonnegative, so that no attenuation occurs in a contrary wind.

23.2.2. Nonlinear, Wave–Wave Interactions

There have been at least two prior attempts to include nonlinear interactions in spectral wave models (Barnett, 1968; Ewing, 1971). These attempts were necessarily crude because only a few theoretical calculations for the transfer had been made at that time. For the present model we represent the nonlinear transfer with a set of empirical orthogonal functions (eof's) that decompose a body of exact calculations for a family of single-peaked spectra with different peakedness and angular spread.

Using a new, more efficient technique, Hasselmann and Hasselmann (1981) have made extensive calculations of the theoretical 2d transfer rates S_{nl} for 18 JONSWAP spectra with peak enhancement factors γ ranging from 1 to 7 and angular spreading s described by $\cos^2\theta$, $\cos^4\theta$, and $\cos^{2p}(\theta/2)$ where p depends on the ratio of wave frequency to peak frequency. We treat these calculations as a data set that can be reconstructed from suitable eof's (Preisendorfer *et al.*, 1981). The Hasselmanns also computed S_{nl} for situations involving windsea and swell interaction. We made no attempt to use these latter computations but their inclusion represents no formal difficulty.

We assembled the set of theoretical transfer rates into a matrix $S_{\text{nl}}(\tilde{f}, \theta, \nu)$ where $\tilde{f} = f/f_{\text{p}}$, and $\nu(\gamma, s)$ is a realization counter with $1 < \nu \leq 18$. The eof representation of these rates, after removal of a mean S_{nl} and appropriate scaling, is given by

$$\hat{S}_{\text{nl}}(f_{\text{p}}, \gamma, s; f, \theta_{\text{II}}) = D(g, \mathscr{E}, f_0)\left[\bar{S}_{\text{nl}}(\tilde{f}, \theta) + \sum_{n=1}^{p'} A_n(\gamma, s) B_n(\tilde{f}, \theta)\right] \tag{23.1}$$

where

$$\bar{S}_{\text{nl}}(\tilde{f}, \theta) = \langle S_{\text{nl}}(\tilde{f}, \theta, \nu)\rangle_\nu \tag{23.2}$$

The eigenvectors $B_n(f, \theta)$ in (23.1) are associated with the covariance matrix

$$C(\tilde{f}, \tilde{f}') = \langle [S_{\text{nl}}(\tilde{f}, \theta, \nu) - \bar{S}_{\text{nl}}(\tilde{f}, \theta)][S_{\text{nl}}(\tilde{f}', \theta, \nu) - \bar{S}_{\text{nl}}(\tilde{f}', \theta)]\rangle_\nu \tag{23.3}$$

The "principal components" of the representation are obtained by "dotting" the eigenvectors with the theoretical estimates of S_{nl}, i.e.,

$$A_n(\gamma, s) = \sum_{f,\theta} B_n(\tilde{f}, \theta)[S_{\text{nl}}(\tilde{f}, \theta, \nu) - \bar{S}_{\text{nl}}(\tilde{f}, \theta)]/D \tag{23.4}$$

where D is given by $a\mathscr{E}^3 f_0^8/g^4$, where a is a constant.

In practical applications most of the elements of (23.1) are computed only once and are stored in tabular form for rapid recall. The variance $\mathscr{E}$, mean frequency f_0, peak frequency f_p, peak enhancement factor γ, and angular spread s are calculated at the end of each time step for the temporally and spatially varying model spectrum. The scale factor D in (23.1) is then computed and the eof amplitude function $A_n(\gamma, s)$ given originally by (23.4) is evaluated from a fifth-degree polynomial that was fitted to the eighteen A_n values in the range $1 < \gamma < 3.3$ and $0.444 < s < 0.699$. Five eof's ($p' = 5$) fit the theoretical transfer well over the entire (γ, s) range and for $0.7 < \tilde{f} < 2.0$, $|\theta_{\text{II}}| < 2\pi/3$. The frequency domain $\tilde{f}$ is tailored to practical wave prediction. Values of S_{nl} for $|\theta_{\text{II}}| > 2\pi/3$ are set equal to zero because they are relatively small and because the noise in the theoretical calculations appears large for angles greater than $\theta_{\text{II}} > |2\pi/3|$. Figure 23.1 shows the approximate and actual transfer at 0 radians associated with spectra at the extremities of the (γ, s) range. The fit is excellent, as

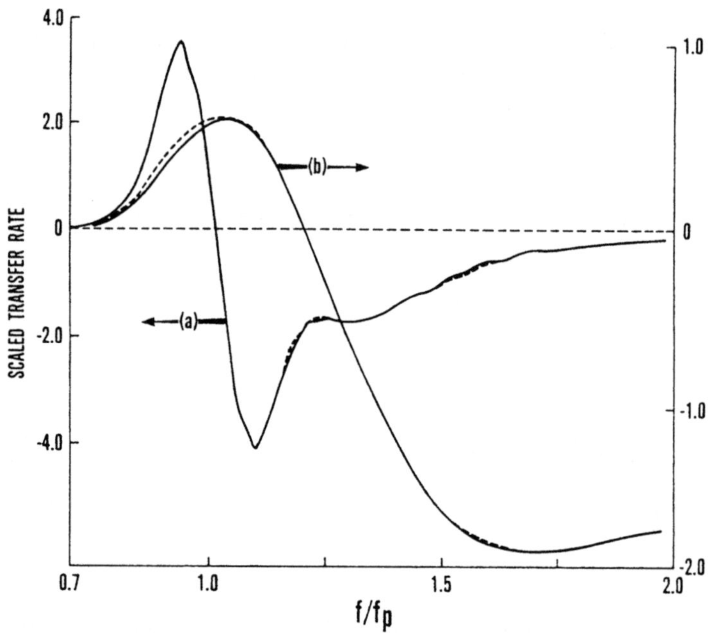

Fig. 23.1. (a) Approximate (——) and theoretical (---) nonlinear transfer for 0° [scaled by $\alpha^3 g^2/(2\pi f_p)^4$] for a JONSWAP spectrum with $\gamma = 3.3$ and angular spreading based on $\cos^{2p}(\theta/2)$ where $p = p(f/f_p)$. (b) Same as (a), except that $\gamma = 1.0$ and spreading is based on $\cos^4\theta$. See Hasselmann *et al.* (1980) for a description of p.

expected. The eof representation also was tested on independent estimates of S_{nl} and was found generally to perform in a satisfactory manner.

23.2.3. Limiting Spectrum

Present wave models rely heavily on a limiting form for wave growth. We employ the well known Pierson–Moskowitz spectrum (E_{PM}) evaluated at the local wind frequency f_U, but with a variable energy level, as our spectral limiter. The energy level is tied directly to the local peak frequency according to the relation found in JONSWAP. The 2d maximum spectral density is taken to be

$$F_{max}(f_U, \theta_I; f, \theta) = \begin{cases} 2\alpha/(0.0081\pi) E_{PM}(f_U, f)\cos^2\theta_I & \text{for } \theta_I < \pi/2 \\ 0 & \text{for } \theta_I > \pi/2 \end{cases}$$

where $\alpha = 0.032\,(Uf_p/g)^{2/3}$, $0.0081 < \alpha < 0.02$. A more complicated angular dependence than $\cos^2$ could be assumed on the basis of recent measurements (see, e.g., D. Hasselmann *et al.*, 1980). Model spectra would then have a more narrow spread near the peak frequency as they approached full development ($f_p = f_U$).

23.3. SOURCE TERM BALANCE

The source terms described in the preceding section must be combined to yield a stable spectrum under a wide variety of wind and wave conditions. We use the minimum number of constraints necessary to achieve this stabilization. One important constraint in the domain of active generation ($f > f_U$) is that the net source function ($S = S_{in} + S_{nl}$) may not be negative. If $S < 0$, then any slight imbalance between the source functions for a particular frequency and/or direction can lead to severely notched spectra to the right of the spectral peak ($f > f_p$). The representation for S_{nl} described in Section 23.2 is based on global spectral parameters and thus cannot respond to any local imbalance in the (f, θ) plane. Compared to JONSWAP spectra, growing spectra from the present model tend to have an f^{-5} slope somewhat closer to the right of the peak frequencies because of the constraint that $S \geq 0$.

A further constraint must be added to handle situations in which the wind field changes rapidly. Common knowledge says that high-frequency waves respond quickly to changes in wind direction, for example, but that low-frequency waves propagate undaunted at large angles to a local wind.

Hasselmann (1963) calculated windsea/swell decay times based on weak interaction theory. In his calculations "swell" was defined as a perturbation spectrum that interacts with an existing windsea spectrum in a certain decay time. He found a more or less hyperbolic relation between decay time and frequency normalized by the frequency of the windsea peak (f/f_p). That is, the decay time was short for $f/f_p \gg 1$, increased markedly near $f/f_p = 1$, and was long for $f/f_p \ll 1$. We imitate the foregoing very crudely by leaving the spectral density unchanged for $f_U < f \leq f_p$ if the density at time level $n-1$ is already greater than the limiting density at time level n. The spectral limiter is applied for $f_U < f_p < f$ whenever the density grows beyond the limiting density over a time step. In effect this procedure causes "swell" at and below the local peak frequency to have infinite decay time, and wave components above the peak to have a zero decay time. For example, if a wind that has been blowing toward 0 switches suddenly toward $\pi/2$, then the spectral density of waves above the local peak frequency must not exceed that of a limiting form with angular dependence $\cos^2(\theta - \pi/2)$.

The preceding logic is formalized in the following way: for $f > f_U$ (active generation),

$$S(f,\theta) = S_{\text{in}}(U;\ f,\theta_{\text{I}}) + \hat{S}_{\text{nl}}(f_p, \gamma, s;\ f, \theta_{\text{II}}) \geq 0$$

if

$$F^n_{\max}(f_U, \theta_{\text{I}};\ f, \theta) < F^n \qquad \text{and} \qquad f > f_p^n$$

then

$$F^n = F^n_{\max}$$

if

$$F^{n-1} > F^n_{\max} \qquad \text{and} \qquad f \leq f_p^n$$

then

$$S \equiv 0$$

for

$$f < f_U \text{(swell)}$$

$$S(f,\theta) = \begin{cases} \hat{S}_{\text{nl}} & \text{or} \\ 0 & \text{if } F^n \geq F^n_{\max} \quad \text{and} \quad \hat{S}_{\text{nl}} > 0 \end{cases}$$

The last constraint in the swell domain prevents wave growth below f_U if the spectrum is already greater than or equal to the limiter.

In practical situations in which it may not be possible to use a relatively fine space–time mesh, we believe it is prudent to make one further simplification to the overall calculations that was not made for the present study. Under a given wind some wave components may reach their limiting values in just a few temporal or spatial steps. These components can be set to F_{max} automatically without going through the spectral balance at all. For example, the JONSWAP fetch relation (for Uf_p/g) could be evaluated for a fetch of one or two spatial steps to give $f_p = f_p(U, x)$. The value of f_p predicted by this relation for a given U then could be used to define a limit above which all frequency components are simply set to their limiting values. Alternately the net growth rates β_{net} measured by Barnett and Wilkerson (1967) could be used to set this upper limit. We find it useful to assume that if $\beta_{net}(f)\,\Delta t > 1$ then wave component f can be set to its limiting value without further calculations. The net growth coefficient has the same form as the β used in S_{in} but is roughly four times larger in magnitude. Either of these procedures has the effect of modeling sub-grid-scale wave growth.

23.4. DESCRIPTION OF NUMERICAL METHOD

The algorithm used for propagation is generally the most complicated aspect of numerical methods in spectral wave models. We use an adaptation of the method of characteristics to avoid the numerical damping and dispersion problems that are typical of finite difference methods. The wave field is quantized as sets of (f, θ) particles moving along parallel rays in given directions and at specified (group) velocities. The rays are thought of as directed line segments that connect equally spaced ray points (Fig. 23.2). The (f, θ) particles hop from ray point to ray point at appropriate multiples of the time step. The zero-influx and no-reflection-boundary conditions are satisfied easily by specifying that incoming particles (at the origin of a ray) have zero spectral density and that outgoing particles (at the end of a ray) are dropped from memory. A net of grid points is defined over the physical domain to introduce the wind field, to accumulate the 2d spectrum values, and to compute source terms (Fig. 23.2). Reciprocal relationships between nearest-neighbor ray and grid points are precomputed and stored. These relationships are the backbone of the overall algorithm.

At time step $n-1$ the 2d wave spectrum at any *grid* point is constructed by accumulating the density values from the nearest *ray* point for *each* frequency and direction. After the spectrum is constructed at all grid points the corresponding source terms are calculated using the formulas of

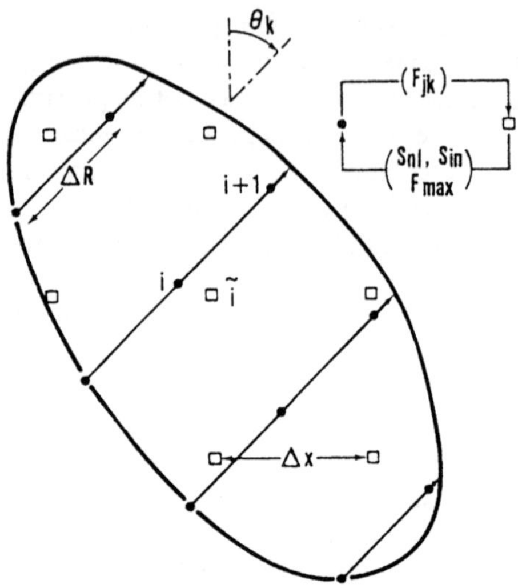

Fig. 23.2. Sketch of very simple ray (•) and grid (□) system. In general a different set of rays is used for each (f_j, θ_k) combination.

the preceding section. Then, to advance to time level n, the spectral density of an (f, θ) particle at any given *ray* point is changed according to the net source function at the nearest *grid* point (with the additional constraints on growth as outlined above). Grid points thus have a region of influence over which the wave field is assumed to be homogeneous. The net of grid points must be laid out to resolve important gradients in the wind and (resulting) wave fields.

Figure 23.2 shows a sketch of a particularly simple ray and grid system. A regular net of grid points with spacing Δx covers an arbitrary domain. A set of parallel rays pointed toward θ_k covers the same domain. The time step Δt and ray point spacing ΔR are chosen such that

$$\Delta t < \left(\Delta R_j 4\pi f_j / g\right)_{\min}$$

$$\Delta R_j = m_j\left(g/4\pi f_j\right)\Delta t, \qquad m_j = m\left(f_j\right), \qquad m = 1, 2, 3, \ldots$$

$$\Delta R_j \lesssim \Delta x \qquad \text{for all } f_j$$

As usual, selection of the time step is keyed to the wave component with the largest group velocity. Particles move every mth step, where m is chosen so that ΔR is about the same at all frequencies.

This procedure, which relies on the local homogeneity assumption, saves considerable computation time. This assumption holds as long as the grid network resolves the dominant spatial variations in the wave field that fall within the frequency range represented in the model. The grid spacing need not be regular. We choose the ray point spacing to be the nearest-neighbor distance in the grid (assuming regularity).

The model equations in discrete form are as follows (for *each* f_j, θ_k particle):

$$\left(F_i^n - F_i^{n-1}\right)/\Delta t = \beta_{\bar{i}}\left(F_i^n + F_i^{n-1}\right)/2 + \left(\hat{S}_{\mathrm{nl}}\right)_{\bar{i}}^n \tag{23.5}$$

and if $\mathrm{mod}(n/m, m) = 0$, then

$$F_{i+1}^n = F_i^n \tag{23.6}$$

where $\bar{i}$ is the nearest *grid* point to ray point i for a given f_j, θ_k. In addition the limits on F between time levels n and $n-1$ are applied as described in Section 23.3. Equation (23.5) applies for a time step when propagation does not occur. Equation (23.6) applies to every mth step, and implies that particles grow then propagate.

23.5. JONSWAP SIMULATION

The discrete nonlinear spectral model as described above was applied to all the ideal cases of wind wave generation designed for the intercomparison study. We describe here only selected results from the case that was designed to test the models against the results of the JONSWAP study (Hasselmann *et al.*, 1973). The physical situation was a 1000×1000-km ocean, initially at rest, subjected to a constant offshore wind of 20 m/s. We remark that the present model was not tuned a priori for this particular situation. A 50-km, regular grid was employed with a ray point spacing of 35–40 km. Twenty discrete frequencies and as many as 24 directions were defined as follows: $f_j = (0.045,\ 0.05, \ldots, 0.08, 0.09, \ldots, 0.16, 0.18, 0.20, 0.225, 0.25)$ Hz and $\theta_k = (0, 15, 30, \ldots, 330, 345)$. Directional resolution of 15° in the downwind half plane and 45° in the back half plane was used. The basic time step was 20 min.

The salient results for this simulation are given in Figs. 23.3 and 23.4, which show variance $\mathscr{E}^*$ and peak frequency f_p^* versus fetch x^* (all variables nondimensional) for successive times during the growth of the spectrum. Good agreement with the JONSWAP relations (plotted in the figures) is found over most of the fetch. Such agreement is a significant

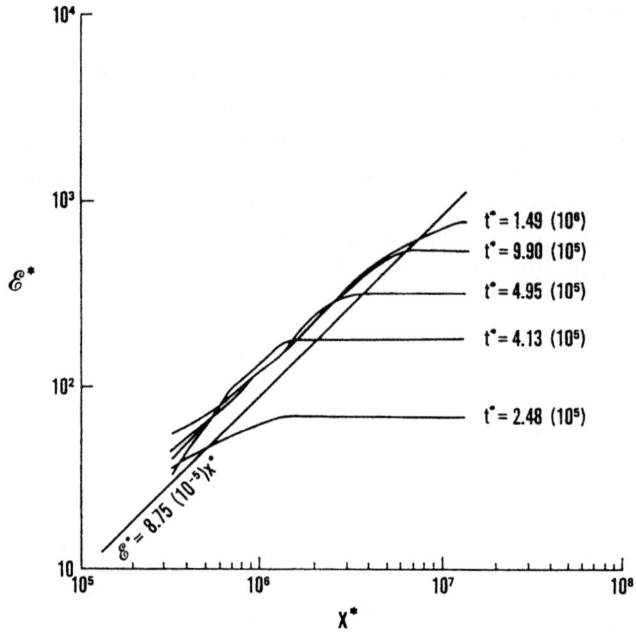

Fig. 23.3. Variance versus fetch for increasing duration (all variables nondimensional), Case II.

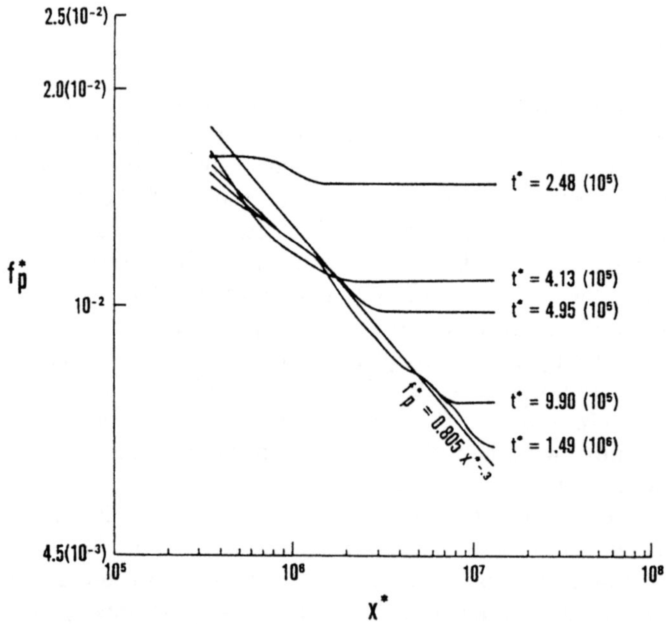

Fig. 23.4. Peak frequency versus fetch for increasing duration (all variables nondimensional), Case II. The PM limit is $f_p^* = 5.6 \times 10^{-3}$.

improvement over previous spectral models. Notably, the model reproduces the JONSWAP relations by combining processes that depend on five independent variables (f, θ, x, y, t) and, again, *without* tuning of the model parameters. Model spectra are still fetch-limited at $x^* = 1.34(10^7)$. A fetch of about $x^* = 1.6(10^7)$ would be required for full development. The model $\mathscr{E}^*$ and f_p^* values scatter somewhat for fetches less than about two grid lengths because the gradients in the wave field are not resolved. The situation is similar to that occurring in discrete hydrodynamical modeling, but not as severe because some wave components actually can grow to their limit in just a few temporal or spatial steps. To reduce the spatial truncation errors we ran a short test on a 200×200-km domain with a 10-km grid and a 10-min time step, and all other model parameters as before. With this fivefold increase in resolution there was no scatter in $\mathscr{E}^*$ or f_p^* at short fetches. In practical cases the difficulty could be corrected altogether by simply setting the high frequencies to their limiting values, as described in Section 23.3.

The times at which the wave field becomes fetch-limited agree well with a simple relation that can be derived from the JONSWAP duration law reported by Hasselmann *et al.* (1976). To use this relation we assume that the fetch-limited domain expands at the group velocity of the peak frequency. Then the extent of the fetch-limited domain X_{FL} as a function of time, starting with no windsea, is found to be

$$X_{FL} = 0.0021\, U_*^2 t^{*(10/7)} \qquad (23.7)$$

using a constant drag coefficient of $1.8(10^{-3})$ to convert to U_*. The times at which $\mathscr{E}^*$ and f_p^* become time-independent in Figs. 23.3 and 23.4 are predicted well by equation (23.7) for $X^* > 1.34(10^6)$. This prediction accurately implies that the present model also follows the JONSWAP duration laws.

The peak enhancement factor of one-dimensional model spectra decreases more or less monotonically from about 2.7 to 1.5 between $X^* = 1.34(10^6)$ and $X^* = 1.34(10^7)$. Wave components greater than about $1.3–1.4 f_p^*$ tend to reach their limiting values as the model spectra develop. So, the high-frequency side of the spectrum follows an f^{-5} slope except in a narrow band to the right of the peak. In this band densities are somewhat higher than for corresponding JONSWAP spectra, perhaps because S_{net} is not allowed to be negative here.

The behavior of the present model in the other ideal cases in the Wave Model Intercomparison Study was generally very good. Model results showed qualitative agreement with existing data and/or physical intuition. Quantitative evaluation of the model against observations, which was not part of the Intercomparison Study, will have to await simulations using real wind fields and measured wave spectra for comparisons.

23.6. SUMMARY

The discrete, nonlinear spectral model described here predicts wave spectra by combining exponential wave growth due to the wind, a compact representation of nonlinear transfer for single-peaked, JONSWAP spectra, and a limiting form based on the Pierson–Moskowitz spectrum. The model is not based on the JONSWAP laws for fetch-limited growth, burt actually imitates them surprisingly well. Additional constraints on the spectral balance are needed to produce stable spectra for general wind and wave conditions. The need to include these constraints may not be a drawback for many practical wave forecasting problems. In order to relax these constraints the representation of physical processes in the model must be improved. Calculating S_{nl} on the basis of local diffusion operators may be one such improvement (Hasselmann and Hasselmann, 1981).

The numerical method used here is an adaptation of the method of characteristics, and has several desirable aspects, as implemented. Numerical damping and dispersion are not present. Irregular geometry can be accommodated, as can variable frequency, direction, and spatial resolution. Finally, the time-consuming parts of the actual code are amenable to vector processing, and computation time could be reduced by a factor of 10 by appropriate reprogramming.

APPENDIX: DEFINITION OF SYMBOLS

(f, θ)	Wave component frequency and direction
(x, y)	Spatial coordinates
g	Acceleration of gravity
t	Time
U	Wind speed (m/s) at 10 m
U_*	Friction speed, 0.855 m/s at $U = 20$ m/s
θ_U	Wind direction
θ_I	$\theta - \theta_U$
$(f_j, \theta_k, x_i, t_n)$	Discrete variables
$F(f, \theta)$	Two-dimensional wave spectrum
$E(f)$	One-dimensional wave spectrum, $\int F d\theta$
$\mathscr{E}$	Variance of sea surface displacement, $\int\int F df d\theta$

f_U	Wind frequency, $0.13g/U$
f_p	Peak frequency of one-dimensional spectrum (determined by quadratic fit in model)
f_0	Mean frequency of spectrum, $\int\int fF\,df\,d\theta/\mathscr{E}$
$\bar{\theta}$	Mean direction of spectrum, $\tan^{-1}\left(\int\int \sin\theta F\,df\,d\theta/\int\int\cos\theta F\,df\,d\theta\right)$
s	Root mean square angular spread of spectrum, $\left(\int\int \theta_{II}^2 F\,df\,d\theta/\mathscr{E}\right)^{1/2}$
θ_{II}	$\theta - \bar{\theta}$
$E_{PM}(\alpha, f_U, f)$	Pierson–Moskowitz spectrum with variable energy level α
F_{max}	Limiting form for spectrum, $(2/\pi)E_{PM}\cos^2(\theta-\theta_U)$
S_{in}	Atmospheric input
β	Growth coefficient in S_{in}
S_{nl}	Theoretical nonlinear transfer
$\hat{S}_{nl}$	Approximate nonlinear transfer
ν	Realization counter for eof's
γ	JONSWAP peakedness parameter, $E(f_p)/E_{PM}(f_p)$
Δx	Grid point spacing
ΔR	Ray point spacing
m_j	Integer number of temporal steps required for propagation of jth frequency component
$\bar{f}$	f/f_p
f^*	$U_* f/g$
t_*	gt/U_*
x^*	gx^2/U_*
$\mathscr{E}^*$	$g^2\mathscr{E}/U_*^4$

ACKNOWLEDGMENTS. We are grateful to Isaac Traxler of Computer Sciences Corporation, who developed the graphics programs and made plots for the model intercomparison. We extend our appreciation to Joyce Ford for typing the manuscript. T.P.B. was assisted by ONR Contract N00014-80-C-0440.

REFERENCES

Barnett, T. P. (1968): On the generation, dissipation and prediction of ocean wind waves. *J. Geophys. Res.* **73**, 513–530.

Barnett, T. P., and J. C. Wilkerson (1967): On the generation of wind waves as inferred from airborne measurements of fetch-limited spectra. *J. Mar. Res.* **25**, 292–328.

Barnett, T. P., C. H. Holland, and P. Yager (1969): A general technique for wind wave prediction with application to the South China Sea. Final report, contract N62306-68-C-0285, U.S. Naval Oceanographic Office, Washington, D.C.

Ewing, J. A. (1971): A numerical wave prediction method for the North Atlantic Ocean. *Dtsch. Hydrogr. Z.* **24,** 241–261.

Hasselmann, D. E., M. Dunckel, and J. A. Ewing (1980): Directional wave spectra observed during JONSWAP 1973. *J. Phys. Oceanogr.* **10,** 1264–1280.

Hasselmann, K. (1963): On the nonlinear energy transfer in a gravity-wave spectrum. 3. Computation of the energy flux and swell–sea interaction for a Neumann spectrum. *J. Fluid Mech.* **15,** 385–398.

Hasselmann, K. (1967): Nonlinear interactions treated by the methods of theoretical physics (with application to the generation of waves by wind). *Proc. R. Soc. London Ser. A* **229,** 77–100.

Hasselmann, K., and S. Hasselmann (1981): A symmetrical method of computing the nonlinear transfer in a gravity wave spectrum. Internal Report, Max Planck Institute of Meteorology, Hamburg.

Hasselmann, K., T. P. Barnett, E. Bouws, H. Carlson, D. E. Cartwright, K. Enke, J. A. Ewing, H. Gienapp, D. E. Hasselmann, P. Kruseman, A. Meerburg, P. Müller, D. J. Olbers, K. Richter, W. Sell, and H. Walden (1973): Measurements of wind-wave growth and swell decay during the Joint North Sea Wave Project (JONSWAP). *Dtsch. Hydrogr. Z., Suppl. A* **8**(12).

Hasselmann, K., D. B. Ross, P. Müller, and W. Sell (1976): A parametric wave prediction model. *J. Phys. Oceanogr.* **6,** 200–228.

Preisendorfer, R., F. Zevlers, and T. P. Barnett (1981): Foundations of principal components selection rules. Scripps Institution of Oceanography, Reference Series, 81-4.

Snyder, R. L., F. W. Dobson, J. A. Elliot, and R. B. Long (1981). Array measurements of atmospheric pressure fluctuations above surface gravity waves. *J. Fluid Mech.* **102,** 1–59.

Chapter 24

The Wave Model EXACT-NL

S. Hasselmann and K. Hasselmann

The evolution of a windsea spectrum is strongly controlled by the nonlinear energy transfer S_{nl} due to resonant wave–wave interactions. Since the exact computation of this process is very time-consuming, most operational wave models are dependent on some form of parametrization of this source function. However, recent, more efficient methods of computing the exact three-dimensional integral expression for S_{nl} (Hasselmann and Hasselmann, 1981, 1985) have made it feasible to carry out integrations of the wave transport equation with the exact form for S_{nl}, at least for simple cases in which only one integration variable occurs.

Three such cases were investigated for the SWAMP study: fetch- and duration-limited wave growth for a uniform wind field (Case II) and a sudden 90° change in the wind direction of an otherwise stationary, uniform wind field (Case VII). A more detailed discussion of the model integrations for these and other cases is given in Hasselmann *et al.* (1985).

In the model energy balance equation

$$\frac{\partial F}{\partial t}(f, \theta; \mathbf{x}, t) + \mathbf{v} \cdot \nabla F = S_{in} + S_{nl} + S_{ds} \qquad (24.1)$$

the input source function S_{in} is taken from the form proposed by Snyder *et al.* (1981) on the basis of direct measurements of the work done by the

S. Hasselmann and K. Hasselmann • Max-Planck-Institut für Meteorologie, 2000 Hamburg 13, Federal Republic of Germany.

atmospheric pressure fluctuations on the waves,

$$S_{in} = \begin{Bmatrix} 0.25 \\ 0 \end{Bmatrix} \frac{\rho_a}{\rho_w} \left(\frac{U \cos \theta_w}{c} - 1 \right) \omega F \qquad \text{for} \left(\frac{U \cos \theta_w}{c} - 1 \right) \begin{Bmatrix} > 0 \\ < 0 \end{Bmatrix} \tag{24.2}$$

where ρ_a and ρ_w are densities of air and water, respectively; θ_w is the angle between the wind vector and the wave propagation direction, $\omega = 2\pi f$; and c is the wave phase velocity.

The dissipation source function was chosen as

$$S_{ds} = -1.2 \times 10^9 \cdot \tilde{\alpha}^6 \omega_p^{-3} \omega^4 F \tag{24.3}$$

where $\omega_p = 2\pi f_p$ is the spectral peak frequency, $\tilde{\alpha} = E\omega_p^4 g^{-2}$, and $E = \int\int F(f, \theta)\, df\, d\theta$ denotes the total wave energy. The parameter $\tilde{\alpha}$ represents a dimensionless shape parameter. For a self-similar spectrum with an f^{-5} high-frequency range, it is proportional to Phillips' constant α.

Since S_{in} is prescribed by measurements (within the limits of the relatively large data scatter), the only freely adjustable source function in equation (24.1) is the dissipation S_{ds}.

The assumed proportionality of S_{ds} to the spectrum follows under rather general assumptions for any nonlinear process which is "weak-in-the-mean" (Hasselmann, 1974). The ω^4 frequency dependence of the proportionality factor in equation (24.3) corresponds to a turbulent eddy-viscosity dissipation, and was chosen empirically to reproduce the observed spectral shape and growth rates.

A relatively high exponent was chosen in the factor $\tilde{\alpha}^6$ to simulate a strongly nonlinear dependence of the dissipation mechanism on the energy level parameter $\tilde{\alpha}$. This characteristic is to be expected if the dissipation is attributed to wave breaking, as is commonly assumed. However, this assumption is not critical. Similar fetch- and duration-limited growth curves were obtained with the factor $\tilde{\alpha}^6$ replaced by $\tilde{\alpha}^2$ and the ω^4 dependence replaced by ω^2 (cf. Komen *et al.*, 1984).

Transition from fetch- or duration-limited wave growth to a quasi-fully developed spectrum was found to depend rather sensitively on the form of the input source function S_{in} near $U \cos \theta_w - c = 0$, i.e., for waves traveling nearly at the same speed as the wind. The frequency dependence of the dissipation source function is also found to have some influence on the transition range. A slightly negative input source function for waves traveling faster than the wind ($U \cos \theta_w - c > 0$), as is frequently assumed in numerical wave models, produces a relatively rapid transition to an almost stationary fully developed state. On the other hand, the combination of zero input for $U \cos \theta_w - c < 0$ and a relatively high exponent in the frequency factor of the dissipation source function, as assumed in equations (24.2) and (24.3), yields a more gradual transition to a "quasi-fully developed" spectrum. In this case the spectrum continues to evolve slowly toward still lower frequencies after the peak frequency f_p has already decreased below the Pierson–Moskowitz peak frequency f_{PM}.

In either case, a strictly stationary, fully developed equilibrium spectrum is never established in the model EXACT-NL. This is because the nonlinear source function S_{nl} continues to transfer energy very slowly from the peak region of the spectrum to lower frequencies, and the energy gained by the low frequencies cannot be balanced by the energy loss due to the quasi-linear dissipation and the (negative or zero) linear input source functions while the low-frequency energy level remains very small. However, the rate at which the weak "energy creep" toward lower frequencies continues in the "fully developed" growth stage depends rather sensitively on the form of the input and dissipation source functions in the neighborhood of the peak.

A more detailed discussion of various choices of S_{ds} is given by Komen *et al.* (1984), who have recently carried out an extended series of experiments with the EXACT-NL model on the energy balance of growing and fully developed windseas. They propose a slightly modified form of S_{ds} proportional to $\tilde{\alpha}^2\omega^2\omega_p^{-1}$.

Numerically, equation (24.1) was integrated using a straightforward first-order forward difference scheme with a dynamically adjusted time step. The spectrum was predicted explicitly in the frequency range $f \leq 3f_p$. For $f > 3f_p$, an f^{-5} high-frequency tail was added (adjusted independently for each θ) to fit the last frequency point which was explicitly predicted. A simple first-order integration scheme was considered adequate, since the size of the time step was determined by the rapidly responding high-frequency region of the spectrum just below the $3f_p$ limit, and was therefore very small in the energetically relevant region of the spectrum near the peak.

REFERENCES

Hasselmann, K. (1974): On the spectral dissipation of ocean waves due to whitecapping. *Boundary-Layer-Meteorol.*, **6**, 107–127.

Hasselmann, S., and K. Hasselmann (1981): A symmetrical method of computing the nonlinear transfer in a gravity wave spectrum. *Hamb. Geophys. Einzelschr.* **52**.

Hasselmann, S., and K. Hasselmann (1985): Computations and parameterizations of the nonlinear energy transfer in a gravity wave spectrum. Part 1: A new method for efficient computations of the exact nonlinear transfer integral. *J. Phys. Oceanogr.* (in press).

Hasselmann, S., K. Hasselmann, J. H. Allender, and T. P. Barnett (1985): Computations and parameterizations of the nonlinear energy transfer in a gravity wave spectrum. Part 2: Parameterizations of the nonlinear energy transfer for application in wave models. *J. Phys. Oceanogr.* (in press).

Komen, G. J., S. Hasselmann, and K. Hasselmann (1984): On the existence of a fully developed windsea spectrum. *J. Phys. Oceanogr.* **14**, 1271–1285.

Snyder, R. L., F. W. Dobson, J. A. Elliot, and R. B. Long (1981): Array measurements of atmospheric pressure fluctuations above surface gravity waves. *J. Fluid Mech.* **102**, 1–59.

Index

GPSR Compliance

The European Union's (EU) General Product Safety Regulation (GPSR) is a set of rules that requires consumer products to be safe and our obligations to ensure this.

If you have any concerns about our products, you can contact us on ProductSafety@springernature.com

In case Publisher is established outside the EU, the EU authorized representative is:

Springer Nature Customer Service Center GmbH
Europaplatz 3
69115 Heidelberg, Germany

Batch number: 10620141

Printed by Printforce, the Netherlands